Skin Theory

Visual Culture and the Postwar Prison Laboratory

Cristina Mejia Visperas

NEW YORK UNIVERSITY PRESS

New York

NEW YORK UNIVERSITY PRESS
New York
www.nyupress.org

References to Internet websites (URLs) were accurate at the time of writing. Neither the author nor New York University Press is responsible for URLs that may have expired or changed since the manuscript was prepared.

Library of Congress Cataloging-in-Publication Data
Names: Mejia Visperas, Cristina, author.
Title: Skin theory : visual culture and the postwar prison laboratory / Cristina Mejia Visperas.
Description: New York : New York University Press, [2022] | Includes bibliographical references and index.
Identifiers: LCCN 2021044596 | ISBN 9781479810772 (hardback) | ISBN 9781479810789 (paperback) | ISBN 9781479810802 (ebook) | ISBN 9781479810819 (ebook other)
Subjects: LCSH: Human experimentation in medicine—Pennsylvania—Philadelphia—History—20th century. | Holmesburg Prison. | Prisoners—Health and hygiene—Pennsylvania—Philadelphia—History—20th century. | Dermatology—Research—Pennsylvania—Philadelphia—History—20th century. | Racism in criminal justice administration—United States—History—20th century. | Racism in medicine—United States—History—20th century.
Classification: LCC R853.H8 M454 2022 | DDC 616.02/7—dc23
LC record available at https://lccn.loc.gov/2021044596

New York University Press books are printed on acid-free paper, and their binding materials are chosen for strength and durability. We strive to use environmentally responsible suppliers and materials to the greatest extent possible in publishing our books.

Manufactured in the United States of America

10 9 8 7 6 5 4 3 2 1

Also available as an e-book

For Manang

CONTENTS

LIST OF FIGURES

Introduction

Science in Captivity

"Here was a golden opportunity to conduct widespread medical tests under perfect control conditions." This was how a 1966 front-page story in the *Philadelphia Bulletin* depicted Holmesburg Prison, which had hosted for the city's leading newspaper a publicity event featuring large-scale skin experiments that were being conducted on its captive population (Katz, 1966). Boasting a prodigious research agenda ranging from the effects of topically applied formulas, like shampoos, ointments, powders, toothpastes, and lotions, to Army studies on jungle rot (athlete's foot), psychotropic drugs, and chemical warfare agents, Holmesburg was home to one of the largest prison research programs of its time, holding contracts with thirty-three drug firms and with the US government, and had become a prominent center for the training of the next generation of notable dermatologists. So advantageous were Holmesburg prisoners to the program's supervisor, Albert Kligman, that the University of Pennsylvania doctor would also found his own privately run pharmaceutical company, Ivy Research, operating solely out of prison grounds. Altogether, these activities, instigated by official invitations to treat prisoners with fungal infections, his area of training, signified a tremendous expansion of Kligman's work at the facility. Kligman had rapidly grown his operations at Holmesburg in the eight years following his first visit, capitalizing on that "golden opportunity" of using caged test subjects to solve with regularity a constant dilemma in medical science research: the making of *perfect* control conditions. At Holmesburg, those conditions meant a captive pool composed mainly of black men (roughly 85 percent to 90 percent of the prison population), about 75 percent of whom were detentioners or were still awaiting trial.[1] Commenting on this subject pool at his publicity event, Kligman remarked, "I feel almost like a scoundrel—like Machiavelli—because of what I can do to them."

Kligman seemingly recognized how heinous it was to use captives as medical test subjects. Nevertheless, he held fast to the virtues of his prison research program, positively identifying with the ethic of ruthless utilitarianism in Niccolò Machiavelli's political treatise, *The Prince*. Infamous for its counsel to rulers that it is much better to be feared than to be loved, Machiavelli's text would also be popularly condensed into the bitter maxim that the ends justify their means.[2] Despite his moral quandary (tepid though it may have been), Kligman perceived in his total authority the optimal settings for scientific discovery, the perfect control conditions enabling his renowned work that would modernize dermatology and the skin-care industry.[3] And his research at Holmesburg was hardly unique. Approximately 80 percent to 85 percent of first-phase clinical trials were being performed in prisons across the country at the time, with the average pharmaceutical corporation conducting at least one of its clinical trials using the penal system.[4] In the postwar United States, experimenting on captive subjects was normal science.[5]

Skin Theory maps these intersecting spaces and representational practices of imprisonment and knowledge production, where the lifeworld of postwar human testing dovetailed with a carceral machinery that had marked captive bodies for disposability and invisibility. Why did the test subject find its ideal figuration in the captive body? In exploring this question, *Skin Theory* does not chart science and captivity as separate processes converging around shared interests or stakeholders here and there in the historical record, avoiding a teleological account of knowledge production in contexts of domination. Instead, it theorizes a common optical rationality between the work of medical science on one hand and the work of captivity on the other. What was it about the prison that appealed to researchers, or what was it about medical science that drew it to imprisoned test subjects, and how might this "what" be traced in the visual artifacts produced from and about medical science research in prisons? Ranging from the documented practice of experts to cultural representations of imprisonment, these artifacts image the scientific structure of racism, a visual logic of racism neither at odds with nor supplementary to the scientific gaze but that in fact can be understood as its cause, its organizing force. That scientific racism abides even now with mountains of demonstrable evidence and bioethical imperatives against it speaks to a scientific rationality internal to racism toward which *Skin Theory* reaches. Thus, *Skin Theory*'s visual- and cultural-studies approach to science in captivity builds on a rich tradition in science and technology studies that reconsiders the social values and responsibilities of science and medicine. In this, the book seeks to uncover the politics, cultural assumptions, and social attitudes embedded within the highly specialized languages of science and technology. In turn, this raises for the field serious ideological questions about the corrective and liberatory promises of knowledge production: Where does knowledge get us in the project of freedom?

Skin Theory comprehends this optical rationality through the skin, the prevailing object of Kligman's experiments at Holmesburg Prison. Skin, Robyn Wiegman (1995) writes, is the visible epistemology of race, "a profound ordering instantiated at the sight of the body" (p. 24). Locating this epistemology in slavery, she argues that "the disciplinary power of race must be read as implicated in both specular and panoptic regimes," functioning as both "the socially inscribed mark of visibility attending spectacle" and the "compulsory visibility attending surveillance" (pp. 38–39). As the visible epistemology of race, skin indexes a gaze that is both scientific and carceral—commanding, reifying, and powerfully constitutive of free and unfree subject positions.[6] The work of *Skin Theory* is twofold, examining how skin constitutes an object of this gaze and how skin also comes to structure it. Reading skin for that which both consolidates and exceeds its representations, the book analyzes how this achievement and incompleteness of vision realizes a routinized antagonism between knowledge and freedom that gives scientific racism its form.[7] Sara Ahmed and Jackie Stacey (2001) demonstrate that thinking through the skin entails thinking with interembodiment, where bodies and worlds touch. Following their call, this book sketches the complicated relationship between visibility and invisibility that captive skin inhered in knowledge production.

In *Skin Theory*, the space of the prison sites the skin as a scientific object and form of seeing. Prison scholars have charted the ways incarceration reproduces older scenes and technologies of black captivity in the United States, such as coffles, slave ships, auction blocks, plantations, and chain-gang cages.[8] As Katherine McKittrick's cultural geography of enslavement illustrates, abolition did not dismantle slavery but instead rendered its built environments a prototype or blueprint for innovating the institution beyond its formal repeal. Mapping the architectures of modern imprisonment to the spatial violence of slavery, McKittrick discerns from within the prison's regimes of "displacement, surveillance, and enforced slow death" (2011, p. 956), a resumption of the "economized and enforced placelessness [that] chained [slaves] to the land" (p. 949). The *where* of blackness, she writes (2006), was rendered *ungeographic* through multiple orders of displacement—displacement from kin, displacement from self—positioning black personhood as having *no place* in traditional geographies and cartographies of the "human." Re-territorializing black bodies from pastoral domains of private ownership (the slave estate) to postindustrial warehouses of public property (the prison), these enduring sites/sights of scientifically managing dispossessed peoples engendered the perfectly controlled environments found in places like Holmesburg Prison, those "golden opportunities" for medical research that made scientific feats like Kligman's possible.

No doubt by *control*, the news story on Kligman's research program intentionally referred to the term's usage in laboratories, wherein any and all factors that may influence test results, called "variables," must be eliminated, minimized, or accounted for. The validity of experimental data hinges on this control of variables, where an effect can be attributed solely to the change or agent introduced by the scientist and nothing else. In fact, the scientist's idealized aim for complete control of variables is that which lends the laboratory its artificiality, able to model real phenomena only by first delimiting them to its tools.[9] As Bruno Latour and Steve Woolgar (1979) write, "Phenomena *are thoroughly constituted* by the material setting of the laboratory" (p. 64, emphasis in the original), allowing them to "escape all circumstances" (p. 239). The laboratory is a culturally unique context and practice in this regard, shutting out much of the world for the meticulous study of something never not in it.[10] Yet the phrase *perfect control conditions* conflates carceral and research spaces, evincing structural continuity between captivity and the making of scientific knowledge, intertwined racial histories well documented in scholarship on scientific racism. Experimentation on slaves and the formative role of captivity and colonialism in modern medicine, as well as historical and ongoing uses of racial categories in medical, scientific, and technological developments, bring into sharp relief how the sciences become modes of racial state and state-sanctioned violence, either explicitly reproducing biological theories of racial difference that legitimize domination or eliding how those assumptions are designed or built into their instruments and procedures.[11] As Dorothy Roberts (2011) argues, the invention of race as an immutable fact of nature remains alongside the sociopolitical orders inaugurated by European colonial expansion and the American slave society, recreated in every modern era to reflect their special forms of subjugation.

On the persistent medical abuse and neglect of black Americans, Roberts and others return to antebellum studies on the enslaved, whose subjection to scientific and medical research fed cultural obsessions with racial essences and biological inferiorities. One of the earliest skin experiments in US medicine, for example, was documented in a slave narrative, a first-person account of captivity told by fugitives and former slaves. In his 1855 memoir, John Brown recounts the excruciating studies devised by a Dr. Hamilton of Jones County, one of which sought to determine the depth of black skin. Only very briefly did Brown describe the study, saying, "This [Hamilton] did by applying blisters to my hands, legs and feet, which bear the scars to this day. . . . He used to blister me at intervals for about two weeks" (p. 48). Hamilton would subject Brown to various other experiments for nine months, ceasing only after Brown's physical condition deteriorated to the point of his becoming unable to complete tasks of hard labor on the plantation where he was kept and from where he would later escape.[12] Hamilton's interest in skin had followed from two centuries of European scientific inquiry on skin color. Thuy Linh Nguyen Tu (2021) describes these theories, which between the seventeenth and eighteenth centuries saw a shift from questions about environment and mutability (do age, climate, or the play of light contribute to these visible markers?) to ones postulating exactly where on the body or in which layer of skin race could be found (do bile, blood, or a special type of mucus contribute to darkening these layers?). Though the advent of microscopy in the nineteenth century identified the cellular basis of skin color (melanocytes), scientists marshaled this new knowledge to further their biological theories of race, ascribing to pigmentation protective powers that ostensibly made colonized and enslaved peoples uniquely suitable for deadly work.

At the end of the eighteenth century, the newly established field of dermatology borrowed these ideas and brought them to novel directions in research concerning the relationship between skin and race, now also turning to depth, texture, and thickness as possible sources of racial differences at the site of the skin. Hamilton's studies on the depth of black skin reflected these new research interests, which resonated with predominant theories privileging surfaces of the body as the primary indicator of race. So entrenched was this epistemological connection between race and skin that emerging anatomical techniques opening up the body's interior spaces to close study remained haunted by racial classifications conjured at the site of the skin. For example, in the nineteenth century, the distinguished American physician and naturalist Samuel Morton published his infamous *Crania Americana*, which posited the larger brain size and superior intellect of whites. His contemporary, American physician Samuel Cartwright, naturalized slavery and pathologized the enslaved's resistance to it, calling the latter a form of mental illness, *drapetomania*, for which Cartwright propounded the psychological benefits of whippings and hard labor.[13] Thus, the size and form of skulls, brains, other organs, and even the interior workings of the mind also became screens through which racial inferiorities could be discerned, corroborating the role of bodily surfaces in unveiling primitive characteristics tucked away in the body, unseen by the naked eye. Skin was, as Samira Kawash (1997) observes, the "sign of something deeper," a sign that "penetrated the body to become the truth of the self" (p. 130).

Early scientific photography sought to record these signs, perhaps most prominent of which were Louis Agassiz's nineteenth-century anthropometric photographs of fifteen enslaved men and women from one South Carolinian plantation. Akin to modern-day mugshots, these commissioned daguerreotypes imaged their subjects from hard front, back, and profile views, stripped naked, some captured in full-body and others from torso-up. Displaying their subjects as brute bodies whose innate differences from whites were ostensibly visibly evident, these photographs, Shawn Michelle Smith (2004) notes, were jarring and disturbing even in their own time, when daguerreotypes were generally used in family portraiture and hence treasured as mementos of loved ones.[14] In contrast, Agassiz's daguerreotypes "viscerally disrupt such expectations; clearly, these images were not made for intimate eyes, but for the cold, hard stare of the laboratory" (p. 47). The plantation setting had transformed the new visual technology from one of personal memory-making into a clinical and impersonal recording of physiognomy. This differential deployment of early photography between white and black subjects demonstrated the racialized standardization of scientific and technical vision in which whiteness figured as unmarked norm, as ideal type, or, in the case of portraiture, as autonomous personhood. Constructed against this norm, the black body imaged the aberrant, the primitive, and the fungible. Agassiz's ideas about race were in line with those of notable peers like Morton and Cartwright. Together, their infamous photographs, measurements, illustrations, and diagnostics reflect one form of scientific racism, where attempts to visibly record and prove inborn differences between races were used to legitimize social, political, and economic hierarchies, or what Nick Mirzoeff (2011) terms the plantation complex of visuality.[15]

Scientific racism in its other form is not reducible to biological determinism. Studies on the enslaved did not always serve to bolster beliefs in intrinsic racial distinctions. Some of these studies instead used slaves as tools or disease models for developing novel treatments, with racial difference constituting a research implement rather than object of study. In her analysis of cultural narratives about pain and suffering, Rebecca Wanzo (2015) revisits J. Marion Sims surgical experiments on enslaved women, only three of whom—Anarcha, Betsey, and Lucy—are named in the historical record. These experiments conducted in the 1850s would earn Sims the prominent title "father of modern gynecology" for inventing both the vaginal speculum and the first techniques for remedying vesicovaginal fistula, a childbirth complication wherein a cavity forms between the bladder and the birth canal. Without anesthesia, the enslaved women endured multiple operations that when perfected, Sims then provided for white patients with the benefit of anesthesia. Calling these experiments "a moment of iconographic resonance in the history of African American women" (p. 154), Wanzo recognizes the complicated and constrained ways we today might see the suffering of Anarcha, Betsey, Lucy, and others subjected to experimental operations where that suffering was invisible—both unthought and categorically denied. In Sims's writings, the technical dimensions of his surgeries and established racialized scientific discourses about pain (that slaves did not feel it), bracketed out any moral consideration about withholding anesthetics and about performing the surgical experiments in general. As Wanzo demonstrates, witnessing the pain of enslaved women and witnessing it retroactively runs up against an archive of expertise where it does not exist and which haunts the continued cultural dismissal of black pain to this day.

Treatments discovered through such dangerous experiments on the enslaved went on to benefit white patients, establishing today's medical culture of asymmetrical care and access to care between white and black patients. Calling these inequities a "medical apartheid," Harriett Washington's (2008) comprehensive history of medical experiments on black people since slavery locates it in the slave's originary status as "medical non-entity," a figure whose practical uses for researchers turned precisely on the inability of slaves to participate in networks of rights and privileges bestowed on recognizable subjects of medical treatment.[16] So although the enslaved could constitute an instrument of experimentation, mobilized for biological theories about race or for generalizable knowledge about human illness, as medical *non-entities*, they could not embody or represent the subject of its achievements. They were chattel, objects of study but not subjects of care, immured in a "dialectic of neglect and surveillance" in which, as Alondra Nelson (2011) observes, the hypervisibility of black bodies simultaneously made them vulnerable to "the constriction of health rights" (pp. 186–187). This captive condition of illegibility-as-usability, founding inequities in care, complicates our understanding of scientific racism. The latter need not seek out essential differences between races. It need not posit racial biology to develop new techniques or understandings. It need not, in short, be about race at all. Whatever its aim, the structure of scientific racism *is* the expediency of racialized bodies, an expediency that endured in the postwar prisoner, whose coerced contributions to medical knowledge, military might, drug discovery, and product safety and effectiveness were hence not a new or outrageous development but a consistent and perhaps anticipated future of antebellum medicine.

Charting this entrenched and evolving relationship between captivity and medical research, scholarship like that of Nelson, Roberts, and Washington at the intersections of black studies and science and technology studies avoids rendering scientific racism a collection of singularly nefarious events caused by misguided research or particularly sadistic scientists. Such works instead convey a more mundane relationship between racial violence and knowledge production, medicine's longstanding tradition of mining the generative capacity of the captive body. In this literature, the ordinariness of scientific racism, rather than its supposed exceptional status, constitutes the object of critical interrogation. For instance, in her historical investigation of raced-based pulmonary medicine in the United States, Lundy Braun (2014) cautions against calling race science a "junk science," since it was, and in some ways remains, mainstream.[17] Race science has exhibited a fantastic resiliency against corrective, "debunking" knowledge, and approaching it as controversy overlooks its deep-seated and distributed influence on expert, legal, and popular imaginings of the body. Similarly, scholarship like that of Smith and Wanzo does not interpret scientific representations of captive subjects as spectacles of abuse, which risks sentimentalizing suffering and sensationalizing the banal violence of scientific racism. Instead, their analyses of the latter's scopic regimes contend with its remarkable tenacity and with the uneasy though no less urgent project of seeing captive subjects differently. Such writings illustrate that hypervisibility and invisibility are not diametrically opposed but intimately linked in their subjectifying effects. Spectacle, moreover, moralizes science by construing it as a neutral or impartial knowledge project—"good" or "bad" depending on its uses, intentions, and outcomes. But rather than portraying science as passive to or determined by social forces and racial ideologies, an approach that takes scientific racism as ordinary and enduring assumes science's active role in the construction of power and its visual terrain.

Skin Theory furthers this analytical approach, taking up human experimentation in prisons not as moral dilemma or spectacle of medical abuse but as normalized, and even celebrated, cross-institutional practices together protracting the long history of science in captivity. It draws its objects from Kligman's scientific papers, as well as from regulatory initiatives and media and aesthetic artifacts later contending with the ethics of his work and of prison research programs in general. These objects point toward the primacy of vision at the junction of science and incarceration within which the captive body formed the object of human and technological perception. However, *Skin Theory* does not aim to historicize or to map the *longue durée* of racial science and medicine to derive empirical or descriptive connections between captive test subjects across different time periods—from plantation "clinics" to postwar prisons. It does not, in other words, create a fuller historical or sociological picture of Kligman's laboratory practice or of prison research programs more broadly, nor does it present the kind of narrative and imagery that might steer readers toward the false impressions and obscuring effects of spectacle. Instead, *Skin Theory* turns its attention to something else pointedly suggested by documented histories of scientific racism: that entanglements between science and captivity were neither accidental nor contingent but *paradigmatic*, that what might be truly sinister about scientific racism is not that it happened but that it happens again and again. This affinity introduces a new predicament in the visual study of science in captivity. For if spectacle is to be abandoned in order to see scientific racism for what it really is, then what it really is, is nothing to see. In *Skin Theory*, what is being opened up for visual analysis and cultural critique is this problem of looking at the generative site of the prison, where forms of *not seeing* were just as pivotal to knowledge creation. Each chapter demonstrates that what remained unseen through the skin did not produce a void in the field of knowledge and experience but was as constitutive of them as were objects of intense scrutiny.

Spectacle allows histories of ordinary violence to forget their ordinariness, ostensibly proffering elaborate detail or richer imagery in the name of more accurate understanding. But if the structural ordinariness of scientific racism is to be grasped, if the automatism of its violence is to be specified, then the question is not one of recuperating or bringing into visibility something that was previously concealed or in the shadows. Neither is it one of refiguring hypervisibility as itself an invisibleizing effect, of enumerating and overturning the unceasing operations of stereotype. In short, the question is not one of representation, as either lack or gratuitous display. Rather, the problem of visibility attending the captive test subject, a medical non-entity or ungeographic figure, reveals a form of seeing and knowing that turns against and effaces its own operations. It suggests that what is unknown or unseen is not a particular historical event or truth but the formidable weight and efficacy of memory. We forget that we remember; we do not see that there is nothing to see—*that* is the representational dilemma of customary violence, of violence on repeat. *Skin Theory* is thus less interested in charting a visual narrative of scientific racism than in studying science as a fundamentally racial optic.

The Laboratory

Contemporary terminology regarding human experiments calls them "clinical trials," which determine the safety and effectiveness of new medical and pharmaceutical interventions. But this book does not call Holmesburg and other prison research programs "clinics" or "clinical sites." It calls Holmesburg a prison laboratory. On one hand, this preferred term more accurately encompasses Kligman's work at Holmesburg. Not all of his experiments tested new drugs or treatments; many performed more basic studies geared toward advancing dermatological research methods (chapter 1). Centering the fundamental workings of skin, disease, and novel approaches of investigation, these studies bore no immediate applications in medicine or consumer product development. Whereas clinical trials view their test subjects as representative of patient populations, the goals of basic research do not address the interests, much less improve the conditions, of their test subjects. It is for this reason that basic research, even in Kligman's time, was not supposed to be conducted on human beings. Generally, animal models were deployed for experiments whose explicit goals included no direct benefits to its test subjects.[18] That Kligman could nonetheless use prisoners for basic research spoke to a principle difference between captive subjects and subjects of clinical trials, a difference that positioned prisoners closer to animals (chapter 3) and that also troubled legal attempts to incorporate prisoners as protected populations in human subject research (chapter 4). *Clinic* evokes the doctor-patient dyad, codifying the test subject as potential recipient of new care interventions. But the dehumanizing practices and spaces of imprisonment troubles this benevolent relationship, and as Kligman's work demonstrates, such practices and spaces make the prisoner a more flexible object of research, more accessible to studies beyond the clinical. Test subjects at Holmesburg were often involved in applied studies, and although *Skin Theory* foregrounds these studies (chapters 1 and 2), it reads them as adaptive uses of prisoners inclusive of but not contained by the term *clinical*.

The Latin roots of *laboratory* link the space for scientific experimentation with the action or process of working, to "labor," highlighting the activities of those designing and doing the studies. Formulating the prison as laboratory site, *Skin Theory* focuses on visual and discursive artifacts produced by those occupying positions of privilege and authority, both formal and informal, with respect to prisoners: scientists (chapters 1 and 2), artists and popular media (chapter 3), and policy makers (chapter 4). Following a science and technology studies tradition of "studying up," *Skin Theory* examines how the different ways these subjects encountered the prison—as site of experimentation, as history of medical abuse, and as ethical problem—exemplify or depict a visual logic that is both scientific and carceral, spatialized in the phrase *perfect control conditions*. Christina Sharpe (2016) calls the prison an ongoing coordinate of the slave ship, a "floating laboratory," now on dry land, still productive of hegemonic claims about race not limited to the scientific (p. 50). As Sharpe suggests, the laboratory that is the prison reflects scientific conventions of looking garnered and practiced through older mechanisms of control and surveillance in captivity. The visuality of the prison laboratory might thus constitute, to borrow from Kimberly Juanita Brown (2015), an "afterimage" of slavery, a problem not of invisibility but of always being seen, always being made visible.[19] Penal control conditions, an extension of slavery's coordinates, were "perfect" insofar as the prison space, like that of the plantation and the slave ship, divided in a Manichean fashion, put under surveillance, and made absolutely accessible for intervention the movements of its captive population—these are the visual arrangements that make for ideal laboratory conditions.[20]

Consider the following 1975 statement of principles by the Pharmaceutical Manufacturers Association (PMA), enjoined by the nation's first federal bioethics taskforce to explain their use of captive test subjects:

> In recent years, as the scientific standards for judging the safety and efficacy of candidate compounds have evolved, it has become increasingly desirable that early clinical trials be conducted *in adequate number of individuals, who are under close supervision for sufficient periods*, so that their responses to drugs can be closely monitored. . . . This has contrib-

uted to an increasing interest in the prisoner volunteer as being *especially suited* for first phase clinical trials and for bioavailability studies of marketed drugs.[21]

Put more succinctly by one PMA spokesperson, prisoners "can't be wandering off to a local beer hall and lousing up your tests" (Wyrick, 1977). Limited forms of escape from wage labor like the local beer hall showed that exploited workers did not make for good test subjects, because even they had too much self-determination. In contrast, prisoners constituted a closely confined and highly managed population, becoming especially suited for medical testing because they materialized the peak control conditions for human experimentation, a ready-made human vivarium. As national regulations imposed more exacting measures for acceptable clinical practice and data, the prison was conveniently positioned to supply researchers with an almost inexhaustible pool of test subjects and the surveillance infrastructure for strict adherence to test protocols. Medical anthropologist Adriana Petryna (2009) writes, "The control over inmates' daily lives, habits, medicinal intake, and food consumption—in contrast to the more chaotic and difficult-to-monitor outside world—would have helped researchers fulfill new mandates for patient uniformity much more proficiently" (p. 63). For the PMA, "few other populations [were] practical or available candidates for these sorts of controlled studies," few other test subjects were "as satisfactory as using prisoners" (ibid.)[22] The prison was a natural laboratory.

That incarceration would be so capacious for human experiments is evident in the rampant use of captive tests subjects in medical science from the 1950s to early 1970s. Author, journalist, and former Communist Party member Jessica Mitford successfully drew public and political attention to this issue in her influential book *Kind and Usual Punishment: The Prison Business* (1973), which remains one of the most comprehensive and politically hard-hitting commentaries on US prison experiments. Prisoners, one physician is quoted as saying, were "cheaper than chimpanzees," a cost-effective means of determining the potency and health risks of new drugs and common household products. Moreover, prisoners were as much a living resource for public scientific enterprises as they were for private ones, also deployed in studies performed by university researchers, public health agencies, and government bodies like the Department of Defense. Examples of major public research studies included those on malaria vaccines, marshaled for military efforts hampered by endemic diseases at war fronts near the tropics; the biological effects of radiation, a pressing dilemma facing the development and proliferation of nuclear technologies in WWII and thereafter; and chemical dependency, an area of behavioral research advanced most crucially in prisons and consequently becoming tied to race- and class-inflected definitions of crime and deviance.[23]

What Melinda Cooper and Catherine Waldby (2014) call the "prison-academic-industrial complex," this nexus of pharmaceutical-, government-, and university-led prison research programs inevitably exploited the wretchedness of prison existence for its steady, replenishable pool of test subjects. A 1976 report by the National Prison Project of the American Civil Liberties Union Foundation (ACLU) determined that all incentives for participating in prison experiments were inexorably shaped through contexts of extreme duress, especially in terms of financial need.[24] Prisoners joined studies to pay for legal expenses like bail, to purchase basic necessities like hygiene products from the prison commissary, and to maintain financial support for family members. Since rehabilitative and jobs programs were scarce in prisons, experiments afforded many more opportunities to work and to break the monotony of prison routine. Their remunerations were also considerably higher at a few hundred dollars a month, compared to the zero or a couple of pennies and dollars a day allotted by other prison industries. Said one test subject at Holmesburg Prison, "I didn't care at the time if the [experiments] could have killed me. I needed the money" (Johnson, 1975).

Furthermore, the noisy, overcrowded, unsanitary, and violent settings of incarceration drove many prisoners to enter the relatively safer spaces of experimental programs. There, prisoners could temporarily ease their fears of assault from both prison guards and other prisoners while also receiving medical attention from programs better staffed and equipped than were the prisons' own clinics.[25] Becoming test subjects, the ACLU argued, provided prisoners these regular though brief respites from the "barren nature" of incarceration, decisions made as desperate attempts to survive its "unbearable" realities. Thus, to the ACLU, the ethical legitimacy of medical research on prisoners was impossible given that "deplorable, oppressive conditions" constituted the "de facto" operations of imprisonment. Indeed, the greater the prison's power over its captive population, the more it seemed to attract researchers. "Most experiments using prisoners," the ACLU noted, were "conducted in medium- or maximum-security institutions, the very institutions where the control is closest and most coercive, the conditions most oppressive and the opportunities for prisoners the fewest."

Though offering opposing views on prison research programs, both the ACLU report and the PMA statement of principles articulated a logical connection between the intrinsically oppressive nature of captivity and the systematicity of observation and intervention in human subject research. But this is not to say that prison spaces and practices automatically bolstered the rigor of research. Allen Hornblum's exposé, *Acres of Skin: Human Experiments at Holmesburg Prison* (1998), the most exhaustive account of the Kligman's research program, emphasizes the rather chaotic, characteristically unscientific aspects of his studies. The prison was a maximum-security facility, and Hornblum described the same circumstances there that the ACLU had detailed in their report on prison experiments. Helmed by a charismatic, well-connected Ivy League doctor able to secure multiple private and public contracts, Holmesburg's research program, Hornblum revealed, was also plagued with disorganized implementation and careless recordkeeping of experiments, whose outcomes were then of questionable validity. Chronicling in harrowing detail the panoply of tests performed on prisoners, Hornblum decried them as both unethical and non-circumspect: Kligman had abused prisoners, taking advantage of their acute deprivation when recruiting them into often unsafe studies; and he had also misused them, carrying out minimally regulated tests that would lead the Food and Drug Administration to briefly pause his research operations because of inaccurate data.[26] Nevertheless, this work would enrich Kligman and distinguish him as a preeminent figure in modern dermatology, helping to transform the science of skin into a mammoth industry of cosmeceuticals even as, Hornblum writes, "[Kligman] churned out so many articles on so many topics that the less credible studies were lost in mountainous verbiage of all the others" (p. 74).[27]

Today, Hornblum's exposé joins many popular and expert accounts defining Kligman's program and prison experiments in general as exemplary of human rights abuses occurring under the auspices of science and medicine. *Skin Theory* moves along the same assumptions, whose reframing of erroneous scientific practice as abuse is an inescapable dimension of political critique within which *Skin Theory*'s thinking remains operative. However, the book also stays with Mitford's cutting observation of postwar prison research programs, about which she had stated at a US Senate meeting, "One researcher told me, 'If prisons were closed down tomorrow, the pharmaceutical companies would be in one hell of a bind.' There is something for everybody in the prison research studies" (United States Congress, 1973, p. 794). The visuality of that something-for-everybody is what *Skin Theory* attempts to bring into relief, asking what it is before us when we look at science in captivity. This is an inquiry that sidesteps issues of professionalization and best practices; for example, whether human experiments should be conducted in prisons. Instead, it rereads what Nicole Fleetwood (2020) calls "carceral visuality" into or alongside the scientific gaze. Fleetwood argues that carceral visuality is both a revealing and a hiding of prisons and of the imprisoned, normalizing mass removal and incapacitation while also limiting the perceptual possibilities of the incarcerated and unincarcerated alike. She writes, "Carceral visuality makes incarcerated people both visible and hypervisible, but also unseeing and unseen" (p. 16). The hypersurveillance of authorities inside the prison is coupled with the invisibility of prisoners to the free world, a field of view bringing to mind the very structure of scientific vision inside the laboratory: the control conditions of experimentation that separate it from the phenomenal world it studies. Linking both visibility and invisibility, seeing and unseeing, in complicated ways, the prison and the laboratory inhabit an optical rationality where the pervasiveness of vision reconstitutes it as absent. In contrast, scholarship on scientific racism has collectively shifted the locus of visibility from instances of recurring malpractice among experts to the very fact of its recurrence, a tenacious reality that their careful historicizing cogently maps out. In this book, skin conceptualizes this problem of visuality introduced by the prison laboratory, whose forms of seeing draw parallels between histories of racist experimental practice and the scientific rationality of racism itself.

Skin Theory

Nina Jablonski (2012), a leading anthropologist of skin color, offers a comprehensive history of skin's privileged role in hundreds of years of racial formation, tracing how colonialism and slavery transformed skin into a "stratifying practice that created identity and hierarchy through social interaction" (p. 141). These looking practices, first generated by purportedly impartial, objective men and machines of natural history, were shot through with moral and aesthetic judgments, where skin became a measure of virtue, intelligence, beauty, and belonging to the human family, as much a measure of character as it was a measure of physical constitution.[28] While this social order is grounded in what is now considered erroneous and backward thinking, its material basis in the body is also quite flimsy, appropriating a space even smaller than that occupied by the whole skin. Skin color resides at a depth of only about 0.05 to 1.5 millimeters, the thickness of the epidermis, whose components include the cells (melanocytes) and cell products (melanin) responsible for skin color. This is a very thin canvas, indeed, for an abstraction as large and enduring as race. One may say, then, that skin has color, but as the violent histories of colonialism and slavery suggest, and to paraphrase Steven Connor (2004), color itself is skin.

The weighty presence of these stratifying practices in the archive of US science and medicine—as drawings, as illustrative text, as photographs—as well as their seminal role in early understandings of the human body and its relation to the natural world, are so manifest of science's ocularcentrism that it may seem counterintuitive to say that they *show* nothing to see. Yet the work of demystifying scientific racism calls precisely for interrogating its evolving normalizations. Each chapter of *Skin Theory* illustrates how race disappears from within the frame to cohere both what stays inside and outside of it, shifting from object of study to the plane of vision itself. Others have theorized the latter in its machine form—race as technology. For Ruha Benjamin (2019), figuring race as technology allows for an abolitionist critique of technology's black box, its exceedingly specialized and inaccessible lingos and practices often presumed objective and progressive. But it is this assumption of neutrality, she argues, that disguises technology's normalization of whiteness, whose status as "invisible center" then magnifies the surveillance of nonwhite bodies.[29] In her rethinking of panopticism, Simone Browne (2015) situates these contemporary surveillance technologies and logics within slavery's regimes of domination, chronicling its racialized management of allowable identities and movements. Biometric technologies, for example, find their earliest forms in slave branding, where "at the scale of the skin, the captive body was made the site of social and economic maneuver through the use of the iron type" (p. 93). Through scarred marks left on skin by hot irons, the slave body publicized its commodity status, its mark of ownership, and in cases of punitive branding, its assumed traits of unruliness.

The concept of race as technology marks the agential capacities of machines, which subordinate racialized bodies and also instrumentalize them to that end. As a technology, race is a *doing* rather than a thing, attending to both visibility and invisibility not only in terms of objects that are hidden or put into view but in terms of diffuse and repeated practice and performance—that is, of material choices conscious and unconscious. Wendy Chun (2009) finds this a useful analytic, shifting the meaning of race from one of essence to one of utility. Focusing not on what race is but on what race does, the technology framework circumvents the nature/culture binary that ossifies race as a passive object of knowledge production, wholly determined by either social norms and traditions or by the innate, unchangeable factors of living organisms. As technology, race can be mobilized differently.[30] This framework shapes some of *Skin Theory*'s approach, which examines how in the experimental setting skin interfaced with a variety of test agents and mediated their harmful effects, becoming a screen through which visualization became inextricable from the construction of racial difference. Chapters 1 and 2 describe how the mechanical uses of skin made it contiguous with other technologies of looking and thus explore where the division between human and nonhuman instruments become muddied or transgressed and therefore impossible to map with certainty.

However, *Skin Theory*'s approach cannot wholly abandon the nature/culture binary given the latter's embeddedness in scientific racism, whose logic the book seeks to excavate. In this sense, it joins other critical scholarship challenging the biological basis of race while also revealing racism's real effects on the body. More importantly still is *Skin Theory*'s focus on the scientific rationality of racism, which the book argues as being nothing to see, precisely because it is constant, unremarkable. *Skin Theory* does not simply aim to reveal and correct flawed assumptions embedded in research questions or in popular representations of race but reads them as general sites of knowledge and cultural production. Ultimately, *Skin Theory* is a text less about technology than about the visual form of reasoning that enables scientific racism to re-place, accumulate, and innovate its tools and messages.[31] To that end, Frantz Fanon's (1952/2008) formidable work on the psychopathology of racism offers invaluable direction. In his *Black Skin, White Masks*, Fanon neither interrogates nor abandons the nature/culture binary on race but resolutely embraces it as his starting point of analysis. For Fanon, the entirety of the binary constitutes the visible terrain of "contact with the white world," and he calls his approach "regressive" because, sig-nificantly, it affirms blackness as "biological cycle" or brute corporeality (pp. 139–143). This forms the basis behind his famous assertion that the "black man has no ontological resistance in the eyes of the white man" (p. 90), refraining from correcting scientific racism to approach it as a form of reasoning that tethers together vision, knowledge, and being in Western epistemology.

One notable way he does this is by comparing the visualizing tech-nique of microscopy to the interpellative force of the white gaze:

> "Look! A Negro!" I came into this world anxious to uncover the meaning of things, my soul desirous to be at the origin of the world, and here I am an object among other objects. Locked in this suffocating reification . . . the Other fixes me with his gaze, his gestures and attitudes, the same way you fix a preparation with a dye. . . . The white gaze, the only valid one, is already dissecting me. I am *fixed*. Once their microtomes are sharpened, the Whites objectively cut sections of my reality. I have been betrayed. I sense, I see in this white gaze that it's the arrival not of a new man, but of a new type of man, a new species. A Negro, in fact! (pp. 89–95, emphasis in the original)

In Fanon's text, microscopy, a technology that renders all parts of the body into surface, negating its volume and shape, encapsulates the gen-eral, white view of the black body, a view that cuts that body into pieces and planes, "spread-eagled, disjointed, redone" (p. 93). *Skin Theory* re-turns to the moral stakes of this oft-cited quote in chapter 1, but what is important to note here is how Fanon actively avoids creating a division of gazes, one belonging to scientists and another to the laypersons he encounters in everyday life, on the street, on the train. Notice the dual meaning of the word *fix* in this quote. The first mention refers to the color chemistry of textiles, or of the dress, decoration, and adornment that Terence Turner (2012) calls "the social skin."[32] Portraying the intractability of meanings attached to the black body, Fanon first likens skin to dye, the stain that locks and suffocates his being into that fantasmatic figure reified and hailed by the phrase, "Look! A Negro!" This is a visual enactment of power through the body that Fanon terms "epidermalization," where corporeal surfaces both sufficiently and totally site/sight white racial fantasies. Bound into this crushing, inescapable specularity in which the white gaze holds unfettered access to and determination of his being, Fanon "discover[s] [his] livery for the first time. It is in fact ugly" (p. 94). It makes him an object among other objects in contradistinction to the Other whose gaze deprives him of his "soul" or of his capacity for meaning-making ("the origin of the world"), rendering him an inert canvas onto which signs could be imputed and transfixed: "cannibalism, backwardness, fetishism, racial stigmas, slave traders, and above all, yes, above all, the grinning *Y a bon Banania*" (p. 92). The initial meaning of *fix* thus addresses skin as mark, where blackness is hypervisible, reified matter.

When read with the description of microscopy, moreover, this meaning acquires added significance, with Fanon's attention to the material processes of scientific vision defining blackness not simply as image but as technique, blurring the boundary between object and subject of the look. Like images seen under the microscope, the black body comprises objectified scientific practice, as constitutive of the gaze as it is its effect. In this white gaze, a *laboratory gaze*, blackness is both content and context of vision, fixed into both the visual products and the material conditions of white ways of seeing. And it is through this scienticizing of the white gaze that Fanon redescribes the procedural nature of everyday looks that transform skin into spectacle. Eschewing the epistemological gap between expert and lay vision, Fanon's historical-racial schema makes "certain laboratories [that] have been researching for a 'denegrification' serum" contiguous with the fearful cries of a white child already deputized in the surveillance of blackness, "*Maman*, look, a Negro; I'm scared!" (p. 91). This historical-racial schema indicates a society and culture that reflects the controlled, remote world of the laboratory—no, smaller still, the minute space between lens and tissue. The white world is a microscope.

However, the second mention of *fix* in the quote troubles this interpretation because this time, skin leaves the field of vision entirely (and the italics in this second *"fixed"* is not an unimportant distinction). What Fanon means in this second reference is not the same as what he meant in the first and in fact even contradicts it; rather than being visualized, blackness is *pre-visual*. For the laboratory technique Fanon describes is more accurately called histology, not microscopy, and though the two are related, Fanon's background in medicine suggests he would have been aware of their important differences: histology is the process by which tissue samples are prepared for viewing with a microscope. Histological methods precede microscopic visualization, and Fanon provides a step-by-step accounting of the former, not the latter. Evincing his medical training, he describes each procedure taken in turn. First, dissection or a biopsy, where small specimens of tissue are excised from the body. Second, fixation, where buffers preserve the dead or dying tissue by preventing it from decaying and thus conserving its shape—here, "fix" is not a coloring technique but a kind of suspended destruction at the cellular level. Lastly, more delicate cutting instruments called microtomes finely section the preserved tissue into thin, translucent slices, each then mounted or pressed onto a glass slide.

This processual description of histology is pivotal to how Fanon defines blackness as lived experience or as "fact." It is a surgical reduction of the body, cut in size and thus abstracted from its whole form. Though not alive, it is not allowed to decompose and disappear but is forcibly maintained in its deadness, like a fossil, petrified. Finally, it is further whittled down, painstakingly carved into nothing but surfaces or nothing but skins. Conspicuously, Fanon skips the last step of this histological procedure, which ends with the staining of tissue samples and their viewing under the microscope. He does not complete the preparation so that it could enter the latter's field of vision. If this step is implied, then the "arrival" of a new human species returns us to blackness as object and technical practice, a discovery made by and culminating in eyepieces squinting at magnified portions of decontextualized body: "The Negro has white teeth; the Negro has big feet; the Negro has a broad chest" (p. 96). If not, or if the opposite is also implied—if Fanon's sense of betrayal occurs *before* the final steps are taken in this mechano-chemical rendering of the white gaze—in short, before histology transitions to microscopy—then the arrival of a new species *precedes the visual*, before the partitioned body becomes representation and thus before the visible and its attendant invisibilities come into being. If the point and conclusion of histology is an image, then this pause in method shifts the focus away from outcomes and processes to their form, to a visual logic that is naught but a stalled process of disfiguration. Fanon's passage does not settle whether the procedure was completed, but the resulting ambiguity in meaning is useful, disrupting what would be a linear, stepwise explanation of race-making. Instead, the arrival of a new species oscillates from visible signs and laboratory practices to something that remains stuck in method, in perpetual preparation and anticipation of its final product, the image.

It is telling, therefore, that Fanon should designate himself as one who waits, alert and attuned to what follows. "If I had to define myself," he says, "I would say I am in expectation" (p. 99). Fanon's theorization of skin as spectacle of race, a hypervisible marker of difference, also contains within it something in advance of the look. This is, in the words of visual culture scholar David Marriott (2007), "a kind of seeing that is both terminable and interminable," a "menace and a postponement of what never arrives and is always about to come," or of what "never arrives and does not stop arriving" (pp. xv–xxi). The spectacle is the usual racist appearance of blackness in the white cultural imaginary that Fanon can count on, ready to detect the familiar always on the verge of materializing, but the waiting itself, the standstill, is unrepresentable—as Fanon describes it, "a zone of nonbeing." Linking expert with popular vision, Fanon locates this zone in both scientific and mass mediated representations, writing "I can't go to the movies without encountering myself. I wait for myself. Just before the film starts, I wait for myself" (p. 119). On this uneasy spectatorship, film scholar Kara Keeling (2003) notes Fanon's "understanding of the role and importance of culture, and of film as privileged cultural form for the production and circulation of the Black imago" (p. 101). Keeling's incisive analysis of the interval just before the film starts, where Fanon prepares himself for an image to come on the screen, illustrates the circular temporality or the "vicious cycle" of black being and white looks.[33] He is both the one who waits and the one he waits for. But outside of this closed circuit of signification, the zone of nonbeing is "an extraordinary sterile and arid region, an incline stripped bare of every essential from which a new genuine departure can emerge" (p. xii). Fanon awaits an image of himself and yet knows beforehand what it will be because it has already arrived, and having arrived, he waits for himself. This is an overdetermination of black being against which an outside, a new genuine departure, appears as impossibility: "A feeling of inferiority? No, a feeling of not existing" (p. 118).

In Keeling's analysis, the interval just before the film starts is no metaphor, Fanon's waiting no figure of speech. Holding immense epistemological import, the interval defines blackness as failed or missing ontology, an absence of change or progress in which, as Keeling (2007) writes elsewhere, "the black is a passing, present preservation of a general past that keeps reappearing" (p. 69). This *is* the structure of white cultural imaginings, not their emblem. I want to argue that in the same way, Fanon's reference to histological techniques is not metaphorical either. Akin to the filmic interval, histology as the protocol preceding microscopy comprises a delayed eventuality of the visible. Fanon's pause in his sequential description of scientific visualization itself constitutes a waiting whose intention is mediated through residual, flattened matter. To illustrate the sameness and repetition into which blackness is locked, Fanon appropriates rather than subverts the optical realism of the sciences. If the white world is a microscope, then Fanon waits for the next discovery that has already been made.

In a text that takes black skin as its title, it is apt that Fanon should narrate histological methods whose objective is to transform volume into surfaces or depth into screens. The connection made to the screen of the film or, more accurately, to the waiting for its display demonstrates Fanon's unique grasp of the nature/culture binary, where both sides (here, science and cinema) come to share the same logic of vision. Together, nature and culture compose the visual dimensions of epidermalization, where the white gaze assumes a dialectical or reversible relationship with skin. Black skin corporealizes this gaze, becoming its mask, a visible collapse or vicious cycle between self and image that Fanon also calls imago. But as Fanon reminds us, epidermalization points not to something weighty, like "essence" or even "interiority," which are the very basis of conflict in the nature/culture dichotomy. Rather, epidermalization is nothing less banal than "experience," or everyday interactions. "I am a slave not to *the 'idea'* others have of me, but to *my appearance*" (p. 95, emphasis added).[34] The titular white mask of Fanon's text may hint at a deeper, hidden meaning or truth of blackness, but the imago illustrates that ultimately, there is nothing behind the mask except another version of it, its repetition in black skin. Race, Alessandra Raengo (2013) argues, "projects a sleeve to the visual, [creating] yet another surface to the image, another skin, an outside layer" (p. 52). In her reading of Fanon, Raengo extends the valence of "fix" to another popular media, photography, whose chemical processes "fixate and fossilize—a Medusa effect" or total exteriorization actualized in and through the imago.[35]

The nature/culture divide may provide opposing accounts of what race is, but challenging or resolving them is not Fanon's concern. His focus is on what engenders the possibility itself of representation. Put differently, race is nature *and* culture, not because nature is culture all along but because the opposition between them marks out the distinct, contingent ways that race can be seen and put to use. Or, conversely, race is *neither* nature *nor* culture, not because of conceptual limits on either side but because their representational capacities or profundity of knowledge claims elide the fundamental emptiness of black being, its absence of ontological resistance.[36] Fanon's theorization of race addresses skin in its broadest sense—the surface of the body and an empty sign—situating it somewhere between corporeality and its abstraction, finding it in the whole conscious and unconscious organization of white society whose contact with blackness is, in experimental terms, one of reproducibility. Hence, it may be more apt to say that blackness does not *precede* the visual but is *of* it. Fanon sees through the white gaze *at* its objects, but he also sees from within it *as* its object, existing as an "image in the third person" (p. 90). A perceptual condition of existing in triple, the imago suggests that the image can look and that it can look back but that this looking proliferates and extends the purview of the white gaze, instrumentalizing the black body in the same gesture that objectifies it.[37]

Addressing skin as apparatus, the first two chapters of *Skin Theory* elaborate on this instrumentalization that establishes a reversible relationship between skin and visualization technologies. They pay close attention to experimental designs, where skin did not simply make up the object of study but the general site of knowledge production, an apparatus of measurement, standardization, and troubleshooting enmeshed with the prison's visual regimes of containment, surveillance, and control. Together, these chapters analyze how scientific vision reconfigures two familiar rituals of racism: the whitening of skin and the dismissal of pain. Yet pace Fanon's approach, *Skin Theory*'s visual archive extends beyond the scientific, also tracing the optical rationality of scientific racism in other objects. This is not to collapse distinct forms of looking but to mark out the visual ambiguity following skin across myriad representations of prison research. Engaging directly with the prison laboratory, the third chapter reckons with its new representational uses, self-conscious efforts recording and aestheticizing it as a vanishing space and thus complicating its dual status as prison and experimental site. The ruins of Holmesburg Prison and their various remediations afford a rethinking of the memory of Kligman's work, which becomes readable not only through his scientific achievements but also through the detritus and decay of his former laboratory. Continuing along these lines, the fourth chapter addresses contexts in which the normalized presence of prison research programs underwent national public scrutiny leading to their significant curtailment under new regulations. Here, *Skin Theory* centers on a different kind of expertise—bioethics, which investigated the extent to which prison populations became administratively available to scientists and doctors. This chapter examines how the prisoner became a "vulnerable" test subject category, elaborating on how their entry into this field of vision recapitulated the prison's role in knowledge production.

Yet in his writings, Fanon relentlessly reaches for what would dismantle the imago. It would be completely novel, a "New Humanism," or as Fanon scholar Lewis Gordon (1995) has described it, a *new science*; on Gordon's reading, Fanon believed Western "science has been much like the history of television. Its producers continue to believe that they bring us the world because they have defined the world so narrowly" (p. 3). But having mapped the optical rationality consolidating the nature/culture binary, Fanon does not proffer a peaceful means of breaking its vicious cycle of representation whether in expert or lay media. Instead, he writes of an explosion (p. 89). In response to fragmentation, Fanon calls for more fragmentation, not less, where the casualties are both whiteness and blackness as such: "The end of the world, by Jove" (p. 191). *Skin Theory* ends with a meditation on the visuality of this explosion.

Resituating the aims of early American bioethics within the political demands of contemporaneous black radical resistance against state violence, it argues against resuscitating the skin's recuperative, intimate, and protective functions, instead proffering both the unreadability and political efficacy of the wound and its proliferation. Abolitionist rhetoric read here unabashedly embraces armed conflict with the state as a legitimate, even necessary, response to the brute force of policing and imprisonment, suggesting a useful reconceptualization of the spectacle of violence. Returning that violence to its creator, this rhetoric seized illegibility for itself, seeking not to protect or heal injury (read: make it knowable) but to rebound it (read: make it spectacular). Where skin has been a material and symbolic force in knowledge production, the wound emerged as a problem for thought by appearing as a problem for appearance.[38]

1

The Skin Apparatus

Seeing Difference

In the midst of Albert Kligman's research program at Holmesburg Prison, Melvin S. Heller, director of psychiatric services at Philadelphia's State Correctional Institution, noted how incarceration itself already constituted a form of human experimentation "involving varying degrees of punishment, deprivation, deterrence, [and] rehabilitation" and put into effect through a "ponderous series of administrative, political, and belated-social manipulations" (1967, p. 24). Providing consultation to the Philadelphia prison system, Heller expressed disapproval of medical science research conducted in carceral settings, stressing in particular the ethical dilemmas of garnering informed consent, that "psychological monstrosity" of human subject experimentation. Heller's critique, however, saw fault not in the prison's adoption of research programs per se but in its failure to fully incorporate the practical demands of experimentation, asserting, "The shame lies not in the fact that it [the prison] is experimental, but that *it is not experimental enough*, either in quality or quantity of effort" (ibid., emphasis added). For Heller, the influx of scientists and technicians into the penal facility highlighted the inefficiencies of prison administration, positioning medical science programs as resources and opportunities for improving carceral management to the benefit of prisoners and custodians alike.

Heller's reflection offers a revealing entry point into the converging logics and operations of incarceration and experimentation during the postwar period. What made the prison space so amenable to laboratory work pointed to a shared ethos of efficiency and control, to two communities of practice wherein the "masters" of calculated injury in one could help guide or illuminate the problems of the other. The scientist and the warden, the technician and the prison guard—each served complementary roles in mobilizing captive bodies for systematic regulation and knowledge production. Still, Heller recognized that medical research in prisons remained an "emotional subject," its controversy issuing from the inevitable problem of informed consent and free will within captive spaces. "Like most monsters," he wrote, "consent may well be a myth. . . . We do things for far more and sometimes quite different reasons than we like to recognize," and he argued that if captive "choice" were to be found at all, it would be someplace between the prisoner and their "psychological milieu," which at Holmesburg Prison comprised an increasingly tangled network of constabulary forces and medical scientific practice.

Before delving further into the ethics of informed consent, that psychological monstrosity vexing bioethicists in the 1970s, the current chapter takes up scientific representations circulating within a network of experts, instruments, and procedures blurring the boundaries between medical research and imprisonment at the site of the captive body. In Kligman's work, photographic images visualized the role of skin as simultaneously an object and instrument of scientific investigation, indexing skin that was both self-referential and reflexively integrated with the methods and tools of research. Thusly mechanized in bringing to light the workings of dermatological science as well as its own, skin was not unlike the photograph itself, composed of a screen and an interface for resolving chemical reactions—a skin apparatus, the scientific gaze made flesh. And as historians Lorraine Daston and Peter Gallison (1992) have noted, this gaze is inextricable from notions of objectivity, the researcher's self-surveillance, or disciplining of personal interests and prejudices. Championed as the mechanical embodiment of objectivity's most cherished value of superior, unflagging self-restraint, the camera was ostensibly the perfect and tireless replacement for biased and unreliable seeing subjects.[1] This gives new meaning to what Heller deemed the lagging experimentality of imprisonment; more panoptic than panopticism, more seeing and knowing than the carceral gaze, the photographic skin/screen pointed to an exceptionally alert and productive functioning of vision making for a power/knowledge, *a scientific method*, that incarceration could only aspire to.[2]

The current chapter and the next turn to images in Kligman's scientific papers, many of which became foundational to the wider field of dermatology. A prolific scholar, authoring and co-authoring over five hundred articles, Kligman published in leading periodicals on skin, including the *Archives of Dermatology*, the *Journal of Investigative Dermatology*, *Contact Dermatitis*, and the *Journal of the American Academy of Dermatology*. His corpus is recognized as having "brought a scientific base to dermatology," its discovery of processes like "photoaging," or accumulated sun damage, and "cosmeceuticals" for treating a variety of skin conditions, "inspir[ing] people to do research in dermatology and to learn about disease."[3] However, scientific publications should not be confused with laboratory practice. Cautioning about the scientific paper's role in obscuring everyday lab work, Karin Knorr-Cetina (1981) writes that it "hides more than it tells on its tame and civilized surface," disclosing none of the messy, behind-the-scenes happenings of laboratory practice (p. 94). Operative in the manufacture of scientific knowledge, published papers elide locally situated selections and decisions made in routine lab work, reconstructing and converting instances of negotiation and compromise into a polished illustration of universality. For this reason, Michael Lynch (1985) argues, photographs in publications function separately and are created for reasons explicitly different from those made in the day-to-day demands of research. Whereas images generated for the lab record and facilitate work as elements in the process of study, those included in articles serve as illustrations of textual material, exemplifying what the article discusses, which is a tidied and narrowed version of technical practice.

Another way to read this formality, or in the case of scientific photography, a morality of prohibition, is through W. J. T. Mitchell's (2005) query, "What do pictures want?," for which Mitchell offers several speculations. Pictures may want to exchange places with the viewer and thus assume the power of the gaze. As objects of stillness and silence, pictures may want to be heard; they may want to be seen or not seen, or they may feel indifferent on the matter. They may want nothing at all, or they may simply want the viewer to ask what it is they want. In all these instances, what pictures want and the power behind this wanting are bound up with what they do not have, what they cannot do, or what they cannot show. What pictures want is not necessarily the same thing as, or is maybe even opposed to, what they mean. To approach Kligman's images beyond their evidentiary functions, or moreover to think of them as desiring objects, is therefore neither to debunk their objective or neutral standing nor to reach for the micro-workings left out by Kligman's publications. It is not to reinstall and intervene in the objectivity-versus-subjectivity debate or to recuperate the prison as the invisible backdrop of research whose operations must then be brought to light through a reinterpretation avowing them inside the pictures, namely by affirming the prison as context and practice of image-making. These forms of analysis, while crucial to demystifying the scientific paper, also sacrifices the latter's essential operation, its unavoidable omission of everyday lab work in communicating to its reading public. This chapter, in contrast, aims to take that omission seriously, countenancing how the interior life of images and their objects constitute precisely what gives them their form as scientific artifacts embodying, channeling, transforming, generating, and trafficking in the social life of science.[4]

Significantly, the scientific paper's indelible erasure of the fullness of lab work creates a topographic association between itself and technical practice that is not at all like a network of people and devices but more approximates the relationship between depth and surface, the paper, like skin, covering up and mediating the insides of the lab. Insofar as that lab was also a prison, the scientific paper then constitutes an important vehicle for exercising and communicating the prison's experimentality to a wider scientific community. In other words, it is the gloss necessarily performed by "tame and civilized surfaces" that effectively situates Holmesburg among other sites of experimentation, curating both in like manner to stage the mutual looking between those inside the lab and those outside it who can only read about it and see it through its images. The scientific paper thus puts Heller's criticism to rest, enacting the prison's experimentality not by revealing differences between laboratory and carceral strategies but by suturing their gazes together.

Screening Difference

Trained in the study of fungi and human fungus infections at the University of Pennsylvania, where he would remain as professor in the campus's school of medicine, Kligman was known to students and colleagues as an exciting teacher and a brilliant and entrepreneurial researcher. He attracted and trained generations of students throughout his career, a teaching legacy felt across the nation and throughout the world, with so-called Kligman disciples becoming faculty and practitioners in dermatology departments in and outside of the United States.[5] Two of note were William L. Epstein and Howard I. Maibach, who, as researchers for the University of California, San Francisco, established their own prison research program with the California Medical Facility at Vacaville. A popular and much-admired scholar in his institution and in his field, Kligman is hailed as having modernized dermatology in groundbreaking work like developing the periodic acid-Schiff stain for the detection of fungi, writing the field's first comprehensive account of the human hair cycle, and characterizing various skin pathophysiologies, including sun damage, contact allergies, and acne, and their possible treatments. Today, prestigious lectureships, professorships, fellowships and other awards bear Kligman's name, though more recent dialogue on his past experiments on minoritized subjects are catalyzing a reexamination of his standing in the history of medicine.[6] Kligman had experience performing medical tests on institutionalized populations prior to founding his program at Holmesburg Prison. Earlier, he had studied ringworm by deliberately infecting the scalp and nails of developmentally disabled children with the fungus, a study that received praise for its "ideal" use of test populations: those in "penitentiaries" (Dr. Frederick Deforest Wideman, as cited in Hornblum, 1998, p. 34). However, Holmesburg Prison would become Kligman's longest-lasting and most sophisticated research site, with up to three-quarters of its captive population participating in medical tests.

Kligman's studies at Holmesburg Prison encompassed an amalgam of applied and basic research endeavors. Basic research sought only to describe or ascertain the biological role of skin, its potential treatment applications secondary to forwarding fundamental theories and methods of dermatology. One example of Kligman's basic research was a study on the skin's permeability, determining to what extent different concentrations of fluorescent dye could penetrate the skin under varying times of exposure.[7] Another outlined a new method for visualizing the skin's outermost layer using light microscopy, a procedure requiring the measured creation and excision of blisters to isolate the desired sheet of skin.[8] Still another and longer study mapped the mechanism behind photoallergic contact dermatitis, a light-induced allergic reaction to photosensitive substances applied to the skin. If a literal comparison could be made between skin and the photographic apparatus, it would find support in these experiments delineating skin as a "persistent light reactor."[9]

In contrast to basic research, applied research was solution oriented. Evaluating the safety of topical agents to be marketed, applied research at Holmesburg consisted of testing the irritancy and allergenicity of lotions, shampoos, soaps, cosmetics, ointments, and antiperspirants supplied by pharmaceutical and hygiene manufacturers like Johnson & Johnson and Procter & Gamble. Applied research also meant exploring possible treatments for skin problems, including an extended study on athlete's foot funded by the US Army, which solicited new therapies for fungal infections afflicting troops deployed in hot and humid climates.[10] However, applied research is not necessarily therapeutic, and indeed, many of Kligman's experiments were not. Studies performed for the military often meant exposing prisoners to infectious diseases and dangerous psycho-chemicals (see coda), and his experiments with dioxin were one of the first to demonstrate the negative health effects of the ingredient in Agent Orange, a powerful herbicide widely used in the guerrilla terrains of the Vietnam War. Commissioned by Dow Chemicals, those studies would eventually be audited by the Environmental Protection Agency and criticized for its use of shockingly high dosages of the toxic substance.[11]

But Kligman's most lucrative applied research project was most assuredly that on the skin-care uses of tretinoin, more commonly referred to as Retin-A, whose multiple patents filed and owned by Kligman point to an array of applications.[12] Belonging to a class of retinoid compounds chemically related to vitamin A, tretinoin is now widely available as both over-the-counter and prescription-grade topical treatments for pimples, wrinkles, and hyperpigmentation such as age spots, acne scars, and freckling. Alongside his third-party work for pharmaceuticals, Kligman's studies on tretinoin helped transform the science of skin into a cosmetic industry. Skin medication alternating as beauty product, or what Kligman would later call a "cosmeceutical," tretinoin's history from research and development to the consumer market shows a strikingly gendered evolution from the skin of black American men behind bars to worldwide sales generally targeting women. Today, tretinoin joins a battery of skin bleaching/lightening regimes collectively valued at $5.8 billion at the time of this writing (with a projected rise of at least another $4 billion by 2027), testifying if not to a global white aesthetic supremacy then to an economic and cultural premium placed onto skin lightness or fairness among self-stylization strategies.[13]

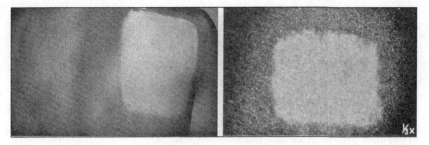

Figure 1.1. (Left) Skin on upper back following six-week treatment with Retin-A. (Right) Skin after nine weeks of treatment. Copyright American Medical Association.

Though also currently prized for its uses in acne medication and anti-aging interventions, tretinoin in Kligman's initial work was used to study neither acne nor wrinkling from sun damage, issues later examined in subjects not in prison. The research at Holmesburg instead prioritized permanent skin discoloration or hyperpigmentation, for which black skin was considered a most befitting model. Investigating problems of "excessive melanization" in whites, Kligman and his team found in black skin the optimal screen for appraising hyperpigmentation and its treatment: depigmentation with tretinoin (Kligman & Willis, 1975). Experiments with tretinoin involved applying the substance to squares of skin on the backs of one hundred black prisoners at strengths comparable to the highest concentrations sold today and in conjunction with two other proven bleaching agents, hydroquinone and a corticosteroid.[14] Exposed to this tretinoin formula twice daily for six to nine weeks, the squares of skin became temporarily depigmented, defined by Kligman as skin color "less than that of fair skinned whites, approaching an ivory hue" (p. 43), becoming distinctly conspicuous against the untreated skin around them (figure 1.1). And for Kligman's team, the darker the skin, the more appreciable the result: "Deeply pigmented blacks were the most susceptible [to the tretinoin formula]. . . . *Obviously*, a 50% reduction in pigmentation will be more apparent in dark than in light skin" (p. 45, emphasis added). More capable of displaying the lightening of skin, indeed of becoming whiter than white, neutralizing all color to approximate that ivory hue, black skin enabled both computational and visual representation of the bleaching process, providing an ideal substrate for showcasing tretinoin's efficacy in achieving a level of epidermal whiteness beyond itself. Administered less frequently (once daily) following "complete" depigmentation, the tretinoin mixture sustained color loss for the duration of study, which in some cases reached as long as six months. Skin began visibly darkening one to two weeks after halting treatment, now exhibiting the reversibility of color changes by progressively rendering "invisible" the exact sites of tretinoin application, a finding that also signaled the presumably benign nature of an experiment that left no traces of itself behind.

Concerned mainly with incidents of hyperpigmentation among white patients, Kligman began his research using a white test subject population, though he quickly found it impractical:

> Initially, white patients with various hyperpigmentation problems were used. This proved tedious and unfeasible. . . . *The patient supply was too limited and too few preparations could be evaluated at a time.* . . . Finally, we hit upon the idea of using the normal black skin of healthy, young, adult, black, male prisoner volunteers. This proved eminently advantageous: depigmentation was easy to appraise, and the effects were repeatable. (p. 43, emphasis added)

In a carceral setting composed mainly of black prisoners, the latter supplied the large sample sizes requisite to replicative, statistically significant data, while their skin enabled relatively "easy" observation of tretinoin's effects. Though dermatologists and physicians usually viewed hyperpigmentation as a superficial or merely cosmetic worry, for Kligman and his team, such changes in skin color could threaten the "psychosocial and psychosexual" identities of white patients and hence warranted serious study. "Pigmentary nonconformists," they said, "are never praised and are generally viewed as odd and unattractive" (p. 40). Their decision to substitute the "normal" skin of black prisoners for medicalized issues in white skin and the "mental anguish" therein betrayed longstanding cultural associations of blackness with abnormality. However, in Kligman's tretinoin studies, black skin was not explicitly pathologized, at least not directly in relation to white skin. White patients did not run the risk of becoming black because of hyperpigmentation issues, so black skin did not figure as a disease of white skin per se. Instead, hyperpigmentation's threat to white psychosocial and psychosexual identity was the loss of *desirable* white skin but not of white skin itself. Rather than an aesthetic benchmark against which white beauty could be compared and contrasted, black skin was instead a biological apparatus for screening differences or complaints *internal* to white skin.[15]

In present contexts of looking, Kligman's photograph of depigmented black skin is disconcerting. Viewers I have encountered express a mixture of incredulity and dismay, most startled into a longer, more intense peering into the photograph as if to assess its authenticity or, more likely, as if their gaze had been caught unawares, unable to turn away from a picture that had forcefully reached out and seized it. The snapshot is gripping—a spectacle of racial difference entrapping the eye to skim over denuded skin of an unnamed prisoner whose back is turned to the camera. Partitioning this captive subject by tightly framing around his shoulders and upper back, the photograph closes in on a scene of scientific discovery to which the prisoner himself was not privy, imputing the scientific gaze with voyeuristic thrill. Roland Barthes (1981) has called the affective intensity of a photograph its "punctum," which arises from the picture to "pierce" or "prick" the viewer, foiling a viewer's attempts at disinterested or leisurely observation.[16] Stupefying or unsettling the viewer with its vivid display of black skin turned white, Kligman's photograph belies the disinterestedness associated with scientific representation. Yet it is important to note that Kligman was wholly unconcerned with locating or analyzing racial differences at the site of the skin. Racial essence did not reach the level of hypothesis in the tretinoin studies, for Kligman did not explain his preferential uses of black skin beyond its expediency. A matter of pragmatics rather than of scientific inquiry, race figured not as research question but as everyday research practice; blackness was utility and utility was its own justification. Scientific methods in general are chosen primarily not for their logical consistency with nature (that they prove something to be true) but for their simplicity and effectiveness in lab work (that they keep research going).[17] Retooled in this way, black skin existed multiply as raw material, as research instrument, as workaround for obstacles, as model for skin conditions, and as visual data, hanging together into a consistent and adaptable medium across its varied uses and thus also sharing with photography the ability to document evidence as well as to function as one.[18]

In Kligman's tretinoin experiments, this meant precisely imaging the dramatic outcomes of experimentation, a deliberate production and presentation of "impressive" results in order to convince the viewer of tretinoin's potency.

> At first, light- and dark-skinned blacks were used indiscriminately. We gradually came to appreciate that deeply pigmented individuals were more susceptible to the depigmenting effect . . . . Deeply pigmented blacks were the most susceptible and whites the least. Lightly pigmented blacks were intermediately susceptible. . . . *We hasten to add that we are not referring here to relative changes.* Obviously, a 50% reduction in pigmentation will be *more apparent* in dark than in light skin. (p. 45, emphasis added)

In other words, Kligman's study was altogether uninterested in mapping tretinoin's activity, or its *relative changes*, in light versus dark skin, with "susceptibility" thus defined not by the test subject's sensitivity or vulnerability to tretinoin's effects but by the viewer's capacity to see those effects. Susceptibility was a methodological concern, a question of visuality separate from biology, with Kligman choosing to use darker skin simply because it made lightened skin look lighter. Formulated as a technical solution to problems of tracking and imaging the course of tretinoin treatment, susceptibility enabled the study's correlation between black skin and hyperpigmentation without need of contemplating or inspecting an organic link between them. Black skin was purely a tableau for seeing white. It was exclusively favored in tretinoin studies for the simple reason that it made possible the "easy" demonstration and presentation of effects, an obvious color line.[19]

By no means the object of knowledge, black skin instead composed what Charles Goodwin (1994) calls "a domain of scrutiny" and "discursive practice" implementing systematic, careful laboratory work, a "professional vision" making visible the phenomenal objects of particular "communities of competent practitioners" (p. 626).[20] In Kligman's tretinoin studies, professional vision meant a conscientious attitude toward refining protocols—adjusting sampling methods to favor darker skin—to more strongly bring out and draw attention to light-dark contrast in the skin. So the picture of depigmented skin is supposed to be jarring, its rhetorical force deriving from a willful and calculated production of epistemologically useful emotion—a burden of representation that is at the same time a burden of feeling.[21] But race did not constitute its subject, the immediacy of seeing race inside the photograph, then, is much like observing a screen without noticing the figures and actions projected onto it. This kind of looking misrecognizes the display for the image by unconsciously registering the presence of something absent within the frame.

The misrecognition of race in Kligman's photograph reflects a profound desire to see that is simultaneously a failure to see, a visual conundrum or "edge of sight" that Shawn Michelle Smith (2013) situates at the heart of photography. Internal to the picture is a thin line between what the viewer sees and does not see, engendering photography's doubled disclosures. Photography reveals, and in this revealing also exposes, all that would remain unseen without its prosthetic capabilities, what is left outside of its frame. And just as the photograph links visibility and invisibility in this way, it also connects the subject and the object of the look by transporting the photographed into the time and space of the viewer. In Kligman's publication, the photograph acts as this spatiotemporal membrane connecting the invasive operations of medical research with the intrusive looking of expert and lay viewers alike. The photograph is didactic, instructing how to visualize the text or, conversely, how to read the image, enacting the *showing seeing* of scientific representations. Misinterpreting the subject of the photograph as one of race therefore speaks to that other operation of photography Smith aligns with its doubled disclosures: showing *not* seeing.[22] Revealing the difference in expertise between scientists and lay viewers, this failed seeing may be ascribed to scientific illiteracy, imputing the photograph with what is not visibly there. Yet as Lacanian psychoanalysis reminds us, misrecognition is not simply blunder, reducible to error, because the difference it creates is productive of self-identity. Whether one sees race or one sees method in the photographs of treated skin, the gaze itself establishes the edge of *site* between racialized subjects and the photographic apparatus.

"Turn White or Disappear"

As with many of his other studies, Kligman's experiments with tretinoin frequently entailed performing skin biopsies for histological analysis,

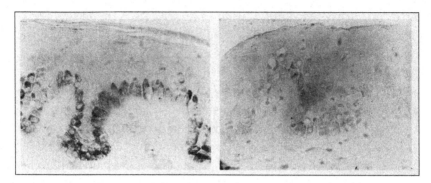

Figure 1.2. Photomicrograph of skin before (left) and after six weeks of treatment.
Copyright American Medical Association.

which deploys microscopes to view cell and tissue structures indiscernible to the naked eye.[23] Following six weeks of tretinoin treatment, pieces of lightened skin were excised from the backs of test subjects and then mounted onto glass slides for observation with a microscope. These slides of skin showed a decrease in the number and quality of pigment granules. Prior to treatment, the latter were more clearly defined by the melanin accumulating around them, suggesting that depigmentation followed from alterations in the regeneration of melanin and in its capacity to protectively coat outer skin cells (figure 1.2). After treatment, the same structures are nearly indiscernible, fading away into an undifferentiated mass of cells. However, treatment did not result in the decolorization or whitening of melanin but in the disruption of melanin's major functions and its production in melanocytes, the cells generating skin pigmentation. In short, less melanin was being produced, and what little was made could not gather around the cutaneous cells they normally would. The latter, researchers found, was due to inflammation.

Treated skin exhibited irritation throughout treatment, with peeling, redness, tenderness, burning, and thickening (acanthosis) peaking at the third week of tretinoin application. While these responses became less observable with time, histological analysis showed a continuous rapid shedding of skin cells at the outermost layers, this accelerated turnover rate contributing to melanin loss via a concomitant loss of cells around which melanin normally disperses. Finally, this more rapid sloughing off of skin was accompanied by the increased presence of white blood cells and of larger, much more active melanocytes, both part of an immuno-logical attempt to heal the skin and defend it from further harm (though the ability of these melanocytes to disseminate melanin was attenuated by tretinoin). Elevated levels of white blood cells and active melanocytes are the body's normal responses to distressed skin, indicating a persis-tent inflammatory reaction in the epidermis. Collectively, the expedited replacement of skin cells and the appearance of more melanocytes and white blood cells confirmed the damaging effects of using Kligman's tretinoin compound at its experimental doses, leading researchers to conclude that "depigmented skin cannot be said to be completely nor-mal" (p. 47). Indeed, the slides are a showcase in the microanatomy of injury.

Photographs taken through a microscope, or photomicrographs, usually require specialized viewers to make sense of them. Without accompanying captions and lengthy explanations about melanin and inflammation, Kligman's before-and-after photomicrographs of depig-mented skin would be unreadable to the non-expert, for whom they may otherwise remain obscure, curious, or perhaps surprisingly beautiful, exhibiting uneven shapes tightly packed together in a single plane, with-out sense of figure or ground. Delineating early twentieth-century labo-ratory visual culture, Lisa Cartwright (1995) finds in photomicrography a Cubist preoccupation with highly abstracted figures, objects stripped of their volume and form and reconstructed into smooth, planar sur-faces merging interior with exterior, or image with screen. In the labo-ratory, this aesthetics of flatness serves a professional vision augmented through instruments and techniques of seeing that unhinge or free it from the body of the researcher and extends its range to shallow spaces normally inaccessible to the unaided eye. Remarking in particular on the ritual torture of animal experimentation, Cartwright illustrates how microscopic vision enters and moves through the internal and infini-tesimal domains of anatomy, how it carves the body into shape, infuses it with the gaze, and further brings it under science's supervision and discipline through ever finer cuts and magnifications.[24] Disembodying both the seer and the seen, photomicrography becomes a powerful ar-biter of experimental knowledge predicated on objectivity's "general ab-horrence" and mistrust of the flesh, its "coolly indifferent presentational style" already containing within it a "narrative of cruelty made all the more stunning by its apparent dullness": an injurious denial of function, complexity, and history, of corporeality, through a gaze that pierces and lodges itself more firmly and painfully within a body's visceral depths (pp. 91–93).

Imbricating skin color loss with skin trauma caused by tretinoin, Kligman's photomicrographs of melanin and inflammation are thus paradoxically more telling of race than are photographs of depigmented squares of skin. Providing nonpictorial, generally unidentifiable representations of a pained body, of a body turned inside out, these highly specialized images evince the troubled relationship between visibility and invisibility, or between recognition and distortion, that is the very operation of race. Dissected, flattened, and scrutinized under the probing lights and compound lenses of microscopy, race disappears at exactly the moment it becomes most clear; we look at it but see something else entirely, like slides of tissue displacing the body for which they purport its hidden meanings. Peculiar shapes and structure—or, to the expert viewer, a tissue's changing cellular composition—these readings of microscopic skin neither reveal nor conceal the truth of race but instead establish its visual language.[25] It is a movement of abstraction, a technically useful misrecognition of a body in pain. In this sense, Kligman's histology epitomizes what Fanon ([1952] 2008) called the "epidermalization of inferiority" or the "imago," with each slide of cutaneous tissue pointing not to an image of the self but to the self as nothing other than an image. What installs the visual exchange between close-up photographs and photomicrographs of whitened skin—or, rather, of black skin under harm—is the dilemma Fanon identified in the psychopathology of racism: "Turn white or disappear."

Writing on dominant views of race in his own time, Fanon alluded to histological practices to delineate the underlying optics shared between reductive and progressive discourses about blackness. Here, again, is his oft-cited passage:

> The white gaze, the only valid one, is already dissecting me. I am *fixed.* Once their microtomes are sharpened, the Whites objectively cut sections of my reality. I have been betrayed. I sense, I see in this white gaze that it's the arrival not of a new man, but of a new type of man, a new species. A Negro, in fact! (p. 95, emphasis his)

Illustrating the searching and surgical quality of the white gaze, a *scientific gaze*, Fanon described how black being was continuously broken down and reassembled into a new object for whiteness, a never-ending process of forcibly making and remaking blackness that Fanon had called a "vicious cycle" (p. 96). Significantly, Fanon located this gaze across seemingly oppositional notions of race, seeing little difference between accounts that posited the biological inferiority of blacks and those that claimed the common humanity of all races. These sentiments, both of which Fanon considered a "vileness," constituted the parameters of knowledge and moral judgment within which blackness inescapably transited back-and-forth as if in a cycle. On one hand, the above passage refers to an antiquated scientific racism, to science's earlier and much more explicit pursuits in racial biology that had included attempts to whiten black skin—a "'denegrification' serum" that would rid black bodies of their "curse" (p. 91). Conspicuously situated within a critique of non-essentialist narratives about race, however, this same passage also takes aim at depoliticized explanations of "color prejudice" (or at what we might today call a post-racial sensibility): "When they like me, they tell me my color has nothing to do with it. When they hate me, they add that it's not because of my color" (ibid.). Such views rankled Fanon just as much and he provided many iterations from his own lived experience:

> "You see, my dear fellow, color prejudice is totally foreign to me." "But do come in, old chap, you won't find any color prejudice here." "Quite so, the Black is just as much a man as we are." "It's not because he's black that he's less intelligent than we are." "I had a Senegalese colleague in the regiment, very smart guy. . . ." "Hey, I'd like you to meet my black friend . . ." "You must understand that I am one of Lyon's biggest fans of black people." (pp. 93–96)

Fanon observes this same condescension in scientific language, finding negligible any differences in sentiment between those experts "rinsing out their test tubes and adjusting their scales in search of that denegrification serum" and those who have proven and agreed that, indeed, "in vivo and in vitro . . . the Negro is a human being—i.e., his heart's on the left side" (p. 99). So whether an empirical, discoverable entity preceding the social world or a purely imaginative fiction immaterial to the natural one, the "fact of blackness" about which Fanon writes marks an evolv-ing, self-correcting white gaze that is as adjustable and reflexive as are visual techniques used in lab work.[26] The photomicrograph itself is so abstract, so unassuming, so obviously technical, and yet just as opera-tive in anti-blackness not because it hides its racism through esoteric imagery—jargon in visual form—but precisely because it is materially demonstrative of the progressive telos of racism. As Fanon recognized with great horror, "And so it is not I who make a meaning for myself, but it is *the meaning that was already there, pre-existing, waiting for me. . . .* I [will] not shape the torch that will burn the world, it is the torch that was already there, waiting for that turn to history" (p. 134, emphasis added). This is the sense of betrayal inherent in his description of histological procedures—that the white gaze finds the same wherever it looks anew.

It is without irony that Kligman's concluding remarks regarding his tretinoin studies should lament a social climate in which his work could be appropriated for shoring up racial hierarchies: "Finally, we per-force must mention one potentially nightmarish outcome of this work, namely, the use of this formula to lighten the skin of normal blacks. We fervently pray that improving social relations will restrain any digni-fied black person from that demoralizing practice" (p. 48). Of course, the experimental setting was excluded from such moralizing, the "po-tentially nightmarish" situation that so worried the researchers exist-ing solely beyond the walls of Holmesburg Prison and not inside of it or because of it. The prison was their laboratory, the captive their free exchange of knowledge, and the "improving social relations" for which they "fervently prayed" therefore standing apart from those systems en-abling their research. In a curious inversion of agency and power, po-tential black patients were entrusted with safeguarding both the ethics of tretinoin use and the dignity of black identity and were called upon by the scientists to exercise restraint against social pressures to whiten their appearance. The personal character and responsible actions of black users, and not those of Kligman's research team, decided whether tretinoin treatment became a "demoralizing practice." Subsumed in this message are some old and familiar themes: a barely disguised fear of racial passing, the exclusion of black people from medical procedures for which they were instrumental in developing, and a preoccupation with science's independence from and transcendence of racist beliefs and attitudes, all of which are embedded inside a dispassionate, scientific language whose unacknowledged foundations in race authorized all the ways blackness could come in contact with it and its creations.

Clinical Subjects

After its capabilities were verified using black skin, Kligman's tretinoin formula was then tested on white patients. Among these, his most impressive results were achieved with melasma, patches of darkened skin typically appearing on the face. All adult white women, these patients reported "great satisfaction" with the formula, which began lightening their patches "earlier than in normal skin of black volunteers." Kligman provided before-and-after pictures of one patient, close-ups of the right side of her forehead showing the striking disappearance of darkened skin from one picture to the next (figure 1.3). Significantly, these photographs are paired with those of black skin after tretinoin application. These latter images do not assume the before-and-after composition that those of the white patient do, suggesting that the viewer sees the "before" category in the surrounding untreated skin framing the areas that had been treated. Whereas progression of treatment must be read *across* photographs of the white patient, this same progression is internal to or encapsulated *within* a single picture of a black test subject. Each photograph of tretinoin treatment on black skin merely duplicates and enhances the visual message in the other. As the gaze moves between white faces and black bodies, the inverted positions of the photographic subjects become apparent. One signals the clinical population of the tretinoin formula, the patient category for whom it was developed, while the other signals the mechanism through which the formula was made, modeling the cosmetic worry shown in the former. There, a white visage becoming less pigmented shows medically relevant data, while in the latter, sections of lightened black skin indicate successful research and development. This contrast also appears in the text of the article, where white patients appear under the heading "Treatments of Disorders of Hyperpigmentation," whereas test subjects fall under "Materials and Methods"—an unambiguous, scientifically significant separation between instruments of experimentation and their potential subjects of care.

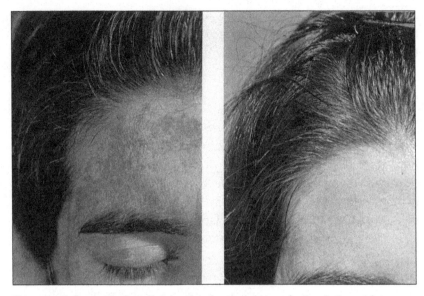

Figure 1.3. Before-and-after photographs of treated skin on forehead. Copyright American Medical Association.

Richard Dyer (1997) has written on film and photography's histori-cal privileging of white subjects, delineating how design and compo-sitional choices in media ensured the illumination and differentiation of white faces, particularly that of women. Early film and photogra-phy, Dyer showed, were aesthetic technologies that controlled and monopolized visible light to make the white figure appear as ethereal, luminous being, whiteness cast as spirit, more-than-body, the subject of enlightenment. A visual medium in Kligman's experiments, black skin was also aesthetic technology but where a different kind of light or lightness was captured or created to maintain the luster of white features. This was brightness produced not by light bulbs but by chem-istry, chasing away shadows in spaces not around the body but right on top of it, in the skin. Visualizing method, the photographs of black skin do not show it to be the opposite of whiteness or even of its con-tamination, blemished skin. They are not, in short, about black skin in itself or as pathological condition. Instead, photographs of black skin define a mode for viewing and reading the photographs of white skin, forming part of the behind-the-scenes laboratory work (the materials and methods) to which the viewer can trace back the medical achievement on display on white brows (the treatments). White skin is hence the subject of all the photographs including the ones featuring black skin, the latter forming its photochemical display, the source of its lightness.

The visual polarity established between photographs of white and black skin does not set them apart as different targets of the tretinoin formula. The photographic quartet does not compare black skin and white skin but brandishes the indisputable potency of Kligman's tretinoin formula in lightening white skin. Evidence of its proof of concept, photographs of black skin visualize the tretinoin formula as property of whiteness and by extension the role of black skin in cohering its aesthetic complaints.[27] As Dyer also argued, the ideological difference between whiteness and blackness is not one of visual opposites, like light versus dark, but of visual disciplining, translucence versus the substance from which it elevates—the radiant white face and the opaque black body, or the subject and its vehicle for (re) presentation.

Kligman's tretinoin compound had also been evaluated in two black patients with vitiligo, a progressive and usually symmetrical loss of skin pigmentation. For Kligman and his research team, this was the only other ethical alternative to using the tretinoin on black skin outside of the experimental setting. "Depigmenting normal skin," they asserted, "is ethically acceptable only in extensive vitiligo" (ibid.), and they instructed patients to apply the tretinoin formula to areas of skin not yet unaffected by vitiligo, in essence to hurry the process of depigmentation along. In this approach, light skin was both the medical complaint and the desired outcome of treatment. Deploying an "obverse technique," "normal" skin became the site of intervention, whereas areas impacted by vitiligo were recast as the new benchmark for successful chemical remediation. The point, paradoxically, was to mask the presence of vitiligo by amplifying its effects. Tellingly, both patients would abruptly disappear from the study, a move that baffled Kligman and his team in light of what they thought were satisfactory responses coming from the patients: "Both seemed pleased. Inexplicably, both defected from treatment; neither could be located for follow-up" (p. 46).

Touching the Subject

All this talk of skin inevitably raises questions about touching and feeling, conceptualizing skin as a zone of intimacy, a place for feeling the Other, or for excavating the haptic qualities of the gaze.[28] Vivian Sobchack's (1992) phenomenology of vision, for instance, theorizes common structures of experience and signification shared between film and viewer, an intersubjective relation wherein the screen sees just as much as it represents, and the viewer, too, expresses while they see. Sobchack calls this dynamic and reversible directionality of looking the "expression of perception" and the "perception of expression," where the eye simultaneously addresses the screen as it absorbs its message, and the screen, too, watches the viewer while presenting its content. Significantly, Sobchack's account of spectatorship explodes the binary between reality and representation by redefining mediation as direct experience in itself. In Laura U. Marks's (2000) work on intercultural cinema, the eye becomes an organ of touch, grazing and spreading out over surfaces instead of delving and searching into their depths. Inviting us to think with the skin, Marks's touch epistemology does not compensate for the limitations of sensorial knowledge, demonstrating in contrast how those limitations generate alternative experiences with media.

Sobchack, Marks, and other feminist scholars' work on embodied vision exhibit a general suspicion of, first, the primacy of cognition in sense-making, and second, the institutional and subjective uses of vision for surveillance, control, and domination.[29] These issues are connected. Abstracting vision and experience from the body denies their situatedness and their capacity for knowledge-formation not rooted in mastery, effecting, in Donna Haraway's (1988) words, a "god trick" that sees "everything from nowhere" (p. 581). Insisting on the sensuous and affective qualities of the gaze, Sobchack, Marks, and others dispel with vision's gendered myth of objectivity—that it is naturally occurring and universal—returning to the body in their assertion that seeing and knowing are always contingent, partial, and intimate. In this respect, Kligman's photographs and the history of scientific representation more broadly typify the kinds of looking that feminist writers warn against, images of objectivity created and imputed with the mechanical detachment and accuracy commonly associated with technology.[30] This

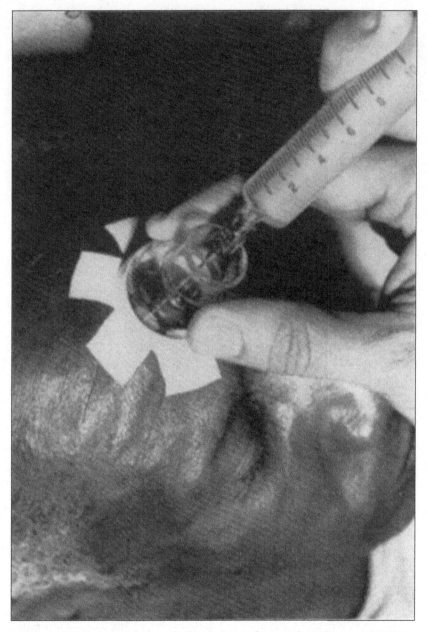

Figure 1.4. Sebum extraction from forehead. Copyright 1958 The Williams and Wilkins Co. Published by Elsevier, Inc.

chapter has intervened in this separation between subjects in possession of the gaze, seeing while they themselves are out of sight, and objects that are possessed by it, trapped in extreme visibility. Firmly located in its instruments and instrumentalized bodies, the scientific gaze decidedly makes visible and material the busy traces of laboratory work.

Still, the skin apparatus examined in this chapter does not easily lend itself to thinking through the intersubjective possibilities of touch, for to think with the skin/screen is to re-encounter the colonizing force of the scientific gaze. In a prison space where physical contact forms a central operation of power, with captive bodies positioned as always already under exposure, touching constrains the perceptual possibilities of the other senses rather than expanding on them, reducing depth into physiognomy and haptics into what Tina Campt (2017) calls "technologies of capture." Kligman's pictures present skin that was and remains open to changing conditions of contact—the eyes and hands of researchers, the analytical viewing of readers, including my own—exemplifying the ways photography can be a "pernicious instrument of knowledge production used to subjugate black subjects" in their present and future (pp. 94–96). This is not to foreclose the possibility of fugitive or reparative readings but to linger on the constrained sensorial and affective lives of such images for which invasive touching was central to their making.

Take, for example, a photograph showing another of Kligman's methods, this time of sebum extraction (figure 1.4). Produced by sebaceous glands, sebum is an oily substance lubricating surrounding hair and skin and was thought to arise through a feedback loop. The sebaceous gland starts and stops discharging sebum in response to the amount of it accumulating on the skin. Kligman hypothesized the obverse, that the output of sebum is continuous or that the glands keep producing more of it no matter how much was already present. To find support for his theory, Kligman measured how much sebum his test subjects were producing over a given period of time, recruiting in particular prisoners who were considered "hyperexcretors" or "sebaceous athletes." One such test subject, an older man, is seen in the photograph showing one of the several ways that Kligman had sought to harvest sebum. The subject sits with his eyes closed and face turned up to the camera, with a small glass cup affixed to the center of his forehead. He had assumed that posture for hours, motionless and supine, after which ether vapor was introduced into and removed from the cup. The photograph shows a pair of disembodied hands using a syringe to inject the gas that would "defat" the area of skin inside the glass cup.

The subject of this photograph, once again, is not the captive. This time, it is the sebaceous gland whose function Kligman sought to better understand. Or more specifically, it is Kligman's *approach* to studying the sebaceous gland that the photograph intentionally conveys. If this image is also gripping, it is because in the present moment and context of looking a critical eye can readily see conventions of representation that divulge the dominated status of the captive. The face, generally known as the seat of identity, is completely unguarded, exposed to the impersonal touching of hands that overshadow it and interact with it only through an intimidating assembly of hypodermic needles and strange gases. To see the captive as the subject of the photograph and not the abstraction he is supposed to be—a human model of the sebaceous gland—can perhaps be considered a kind of touching, one that reaches for his individuality and affirms the viewer's rightful sense of alarm and unease belied by the utterly clinical presentation of what is happening to his face. But this seeing is also enabled by the photograph, which brings the viewer so very close to the captive, almost as near to his face as are the steady fingers manipulating the syringe held tightly to his brow. Looking down onto his placid expression, the viewer assumes the position of the one who took the picture as well as the one performing the injection—shooting into the subject through both the camera and the needle. This photograph is like any other among Kligman's tightly framed images of skin, photomicrographs and close-up photographs, exemplifying the "shattering" of the individual that Marks also identifies in the more malevolent undertones of touch epistemology. What brings the viewer into intimate contact with the captive body is that which also renders it scientifically useful: we see, but we touch an abstraction.

Extended across laboratory researchers, visual instruments, notebooks, and publications and their audiences, scientific representation, as Michael Lynch (1998) described it, is an "externalized retina." This is neither a furtive nor an untouchable eye but one whose augmented powers of seeing issue precisely from its ability to bring more bodies, objects, and tools nearer to each other. Although the photographs discussed in this chapter do not illustrate exactly what happened in Kligman's work at

Holmesburg, they do mediate how that work travels beyond the prison-laboratory to a wider scientific community. As Lynch (1988) has also argued, illustrations in scientific publications function as displays, accomplishing a consensual seeing between the creators and viewers of an image or between the writers and readers of a paper, this shared optics staging their shared competency. Thus irreducible to data or to their referent, scientific representations also materialize vision as distributed practice and cognition between instruments and practitioners inside the lab and between experts and stakeholders outside of it. They are social achievements, contingently produced and selected for their capacity to inscribe method. And they achieve sociality, intentionally made for or responding to disciplinary conventions of scientific representation.

In Kligman's publications, this interactional seeing that connects instrument, researcher, and reader transpires across seemingly neutral or dispassionate images, which convert the affective and sensorial dimensions of close-up photographs and photomicrographs into something much less intimate and more clinical. Bringing the viewer nearer to the viewed object, Kligman's pictures of skin do create impressions of proximity or closeness, imbuing the gaze with tactility by transporting it to, or almost in the same plane as, the cutaneous surface over which it alights. Yet skin experimented on unsettles the notion that such contact is always mutual or relational—that to touch is at once to be touched, and the eye is also seen and caressed by the image—instead demonstrating a penetrative and abstracting gaze that creates a sense of deep space, of objectivity, through intimacy imputed with the probing attentions of scientific observation and manipulation. Pulling the viewer into the image and allowing the eye to commingle with skin, Kligman's photographs do not so much transform looking into feeling as they do touching into the ocular centrism of science, which does not erase the body but re-incarnates it as gaze.

2

Skin Problems

Seeing Pain

In the last chapter, the skin apparatus conceptualized how captive bodies became instruments of visualization, where race constituted the mode rather than the object of seeing. The current chapter addresses what happens to race when it does become the object of study, when the skin apparatus becomes explicitly racialized in scientific inquiry. While the last chapter foregrounded the capaciousness of skin, this chapter centers the politics of naming that agency, shifting from issues of knowledge production to the production of pain. In his hyperpigmentation studies discussed in the last chapter, Kligman did not hypothesize or seek to examine the condition in black skin. Melasma, acne scars, and liver spots—these, Kligman thought, were "mainly a problem of whites." Not until he noticed the scarred and blotchy skin of his black test subjects, including those who were involved in his skin lightening studies, did Kligman become more attuned to the condition arising in black skin. Across skin tones, dermal injuries like sun damage and allergic reactions can cause affected areas to produce more melanin, a phenomenon toward which Kligman's "own experience with blacks"—by which he meant his experiments on prisoners at Holmesburg—now "awakened" his sensibilities. In his skin-lightening experiments with tretinoin, for example, Kligman found that following cessation of treatment, the skin of several test subjects exhibited higher levels of pigmentation than they had prior, attesting to the adverse effects of the formula at experimental doses. He observed the same outcomes in many of his other skin experiments, where repeated exposure to chemical and physical trauma regularly produced "intensified" pigmentation where skin injury was sustained.

This admission of the enduring effects of his experiments, clearly acknowledged in his scientific papers, stands in contrast to his vehement public denials of the excessive risks and lasting consequences of his studies. Throughout his career, Kligman insisted on the safety of his experiments at Holmesburg, once commenting, "To the best of my knowledge, the result of those experiments advanced our knowledge of the pathogenesis of skin disease, and no long-term harm was done to any person who voluntarily participated in the research program."[1] But as one former prison guard noted decades after Kligman's program ended at Holmesburg, "If you ever saw the guys [test subjects] on the beach, you would know where the hell they've been" (Kinslow, as cited in Hornblum, 1998, p. 10). These test subjects were marked for life, their disfigurement a compulsory disclosure of their past confinement at Holmesburg, linking them to captivity from the outside while also making corporeally visible the truth ambiguously disavowed in Kligman's words. Stigmatized, bearing what Kligman called the "dermal tattoos" of his experiments, they carried on their bodily surfaces imprints of the prison laboratory, or they were, to paraphrase Jay Prosser (2001), "re-membered" only as skin. Thus, Kligman's discussion of hyperpigmentation in the aftermath of continuous experimental harm did not amount to a moral recognition of injury. Rather, for Kligman, hyperpigmentation among his test subjects signaled yet another potential subject pool for testing or applying his skin lightening formula, using their injuries from earlier experiments to propose expanding the uses of his new tretinoin treatment. In black test subjects, Kligman wrote, "the most trivial chemical and physical traumata, frequently unnoticed or unrecollectable, tend to produce persistent hyperpigmentation" (p. 43), which could hence be targeted for his formula's whitening effects. These test subjects were now readable as patients and consumers, or at least as clinical subjects, through wounded skin first caused by research and development, an instance of serendipity or unexpected fortune common in the experimental setting.

Kligman's statement articulates his ambivalent recognition of black pain, a pain going "unnoticed and unrecollectable" until it became operationalizable for other studies—clinically relevant, generating new questions and leveraged for new applications. And this instrumentalized definition of black pain illustrated the contradictory nature of its scientific importance, where it was concurrently "trivial" and inconspicuous but also "persistent" and unmistakable. The current chapter attends to this troubled legibility of black pain at the site of the skin, turning to Kligman's earlier studies, whose intentional and constructive production of pain was, despite his claims, in no way innocuous. That pain and its complaints would have a racial politics, Keith Wailoo (2014) argues, is unsurprising given the nation's long history of denying black suffering and centering white anguish since slavery, explaining ongoing racial disparities among both recipients and givers of care. Identity continues to inform the recognition and treatment of pain, and dermatology is no exception to this tradition. Today, delayed diagnoses due to inadequate training among practitioners and structural inequities in access to care has resulted in higher morbidity and mortality rates in skin cancer among black patients as compared to white patients. The latter are also more likely to receive therapeutics for other skin conditions that range from mild to severe, contributing to greater disease severity among black patients even as the latter are also more exposed to environmental pollutants harmful to the skin. This has led to several calls for a more diverse field of practitioners that remains predominantly white and for training more inclusive of darker skin tones, which, too, remain underrepresented in dermatology textbooks and guides.[2] Invisibleizing the injuries they inflicted on black skin, Kligman's prison experiments reflected and prolonged these ideological disparities in relief with which dermatology now reckons.

The studies discussed in this chapter aimed to generate a reliable measure of allergic reactions in the skin, one that hinged on visible markers of injury. Though not encapsulating all of Kligman's research on skin allergies, these studies unequivocally differentiate black skin from white and contribute to an area of dermatology—dermatitis—for which Kligman's work is also recognized as foundational. Race, this chapter will show, figured explicitly in the optics of pain that Kligman sought to standardize. Kligman found himself accounting for race in ways he did not have to in his tretinoin experiments. In the latter, black skin composed a mechanism for exploring treatments for hyperpigmentation, a backdrop for observing the process of skin becoming white. As an instrument used for studying something else, black skin did not require an explanation for use beyond its convenient availability. In his studies on allergy, however, Kligman's preferential deployment of black skin became a methodological problem, a variable that got in the way of research and therefore called for meticulous troubleshooting. For scientists, black skin frustrated their efforts to observe and report the results of their work, obstructing their very object of study: injury. Whereas the skin of black prisoners constituted a favorable medium for seeing white, when it came to making and viewing images of pain, black skin hampered perception. This impelled Kligman to then investigate the epistemological significance of his most common instrument of visualization: black skin, now unruly, resisting the uses to which it was called on to perform.

Posthumanist literature in science and technology studies might call this matter's form of agency "thing power " (using Jane Bennett's [2010] term), when the breakdown of devices and material infrastructures reveal most profoundly a persistent and active force making up the agential capacities of matter and their recalcitrance to social construction, human designs and representations. It is when objects malfunction or behave unexpectedly that their human makers and users can no longer take their normal(ized) functions for granted.[3] Alternatively, posthumanist scholarship that more centers the body might better situate the problem of black skin in Kligman's studies on pain, also revealing the immanent vitality of corporeal or biological substances irreducible to the self-knowing, self-possessed human subject they make up. Advancing from feminism's longstanding recuperation of "the body" against Cartesian dualisms subjugating it to the masculine sphere of mind and spirit, posthumanist theorizations about the physical stuff of organic matter would provide a compelling interpretation of Kligman's work, where the cerebral activities and creations of scientists become inseparable from the uncanny, cyborgian bodies of their human implements.[4]

Reading captive experimentation through these often poignantly affirmative theories of nonhuman agency is awkward. Their recovery and defense of the "missing masses" (matter) is difficult to square with the dehumanizing practices of imprisonment and scientific exploitation.[5] How can one affectively and theoretically reconcile the experimental life of the sciences with the social death of captivity? Yet Kligman's research program, among many others, forces such a reading, where the materiality of violence complicates posthumanism's favorable recasting of the human (which is to say, *all* humans) in non- or inhuman terms. If to insist on the agency of matter is to decenter the human subject, then what is this agency in instrumentalized captive bodies, a displaced humanity installing the liberal humanism against which posthumanism makes its interventions? As theorized in the work of sociologist Orlando Patterson (1982), the deathness of racial captivity issues from the captive's limitless availability to brutality, a violence both material and symbolic whose condition of being without limits renders it unrecognizable as violence.[6] This violence is not punitive but everyday, and it is also not contingent but awesome in its generativity. Zakiyyah Jackson (2020) calls the latter "plasticity," "a mode of transmogrification whereby the fleshy being of blackness is experimented with as if it were infinitely malleable lexical and biological matter" (p. 3). In Kligman's experiments at Holmesburg Prison, what black skin was made to show and do conveyed this creative plenitude, which tethered the practice and meaning of agency—an agency that was transient and pained and, significantly, mediated through scientific representation. Also underscoring the production and exchange of knowledge that bondage made possible, Harriet Washington described involuntary experimentation as the "personification" of captivity, suggesting a posthuman that is not altogether different from the one materialist science and technology scholars often situate in more contemporary ethical and political discourses about an increasingly techno-mediated world. The uncanny thing-ness of bodies and the equally uncanny liveliness of things—this marked the troubled agency of skin and image made more animate than the captive subject they both visualized.

The question of pain and its legibility helps ground this cumbersome reading of agency in Kligman's allergy experiments, where the planned and deliberate inflicting of pain became less effectual than the body's perceived obstinacy in showing it. Skin color mattered to the recognition or, more accurately, to the realization of pain, objectifying it as a phenomenon that can be scientifically studied and manipulated in a prison setting where death beyond the biological and life beyond the social met. Elaine Scarry's (1985) thoughtful analysis on pain, representation, and the connective role of the apparatus illustrates how agency escapes individual persons and things, sometimes projected onto the suffering body to render its pain invisible, sometimes actualized in the tools or weapons used to produce pain, and at other times emerging as the language that makes one's pain real to others able to intervene in and redress pain or, conversely, that makes pain the property of the powerful who mete it out. This suggests that posthumanism's emphasis on agency and liveliness might be deflecting passivity and inertness as significant dimensions of pained material and human being. As Claire Colebrook (2018) argues, while posthumanism insists on the complexity and multiplicity of matter's being/becoming, less remarked upon is the complexity and multiplicity of deathness, as if stasis or incapacitation can be easily enough understood. The current chapter addresses pain and representations of pain as this complicated, multiple, and dispersed expression of captive agency, a precarious but no less potent form of material being inextricable from scientific method in the prison laboratory.

The Maximization Test

With many of his research contracts focusing on the irritancy and allergic potential of commercially produced personal care items, Kligman was acutely attentive to the procedures and tools he used to establish the safety of these products. Patch testing was one such approach, a diagnostic tool first developed at the turn of the twentieth century to help diagnose skin sensitivity to certain substances. Now more commonly used to identify allergies, the procedure generally involves applying test compounds to patches of fabric and then taping those patches onto the patient's arm or back, sometimes arranged into rows and columns. After a few days of exposure, the panels of patch tests are visually inspected for any signs of allergic reaction: lesions, rashes, welts, swelling, reddening, and itching. These same symptoms, however, also appear in irritated skin, which is clinically different from allergic reactions. Irritants physically damage and therefore directly impair the skin's protective outer layers, whereas allergens prime the immune system to mount an inflammatory response. The former refers to compounds directly acting on the skin (like solvents in cleaning products), the latter to conditioned immunological reactions occurring in predisposed persons (like latex or nickel allergies). Studying both irritancy and allergenicity, Kligman was keen on distinguishing between different medical phenomena presenting similar symptoms and sometimes precipitated by the same agent.

The patch test constituted Kligman's primary mode of evaluating toiletries and cosmetics to be marketed to the public, but he became concerned with enhancing the method's accuracy at a time when weak allergens and irritants were becoming more prevalent in common household goods. These substances were considered weak because they did not produce noticeable symptoms after prolonged use. Upgrading, and in some instances replacing, the patch test to better detect the effects of these milder compounds, Kligman's new bioassays indicated his shifting objects of study, from questions about safe levels of topically applied chemicals to the epistemological significance of his instruments, including skin. Experimental instruments, Karen Barad (2007) writes, set apart measured objects from measuring agents and unambiguously determine the properties or elements of what is being measured. However, this boundary produced between cause and effect, or observer and observed, becomes blurred when the apparatus itself comes under scrutiny, taken as a phenomenon of study whose inscriptions and mediations then acquire reflexive importance. In such studies, skin is not a passive site of observation or manipulation but is consciously recognized as an active agent whose powers of visualization Kligman sought to better understand and master.

One major breakthrough was dubbed the "maximization test," which structurally did not differ much from the patch test, consisting of small squares of absorbent, non-woven cloth pressed onto the skin and dressed with overlapping layers of plastic tape, the patches measuring from one centimeter to one and a half inches.[7] Kligman's images of the maximization test display its final construction, exemplified in a series of three black-and-white photographs (figure 2.1). The first two show the same patch securely fastened and sealed onto the inner forearm, differentiated by the more liberal addition of tape in the second. Created to more accurately measure the potency of allergens, the optimized sensitivity of this construction was achieved through a compendium of experiments determining, for example, the kind of patches and tapes to be used, the different concentrations of weak and strong agents to be administered, the duration and number of chemical exposures per subject, which extremities or areas of the body should be tested, how large and how many patches should be applied to a single test subject, and what medium or vehicle—water or petroleum jelly—could dissolve and carry each test agent to the application site. For such a wide constellation of experiments, each one replicated multiple times to achieve reliable

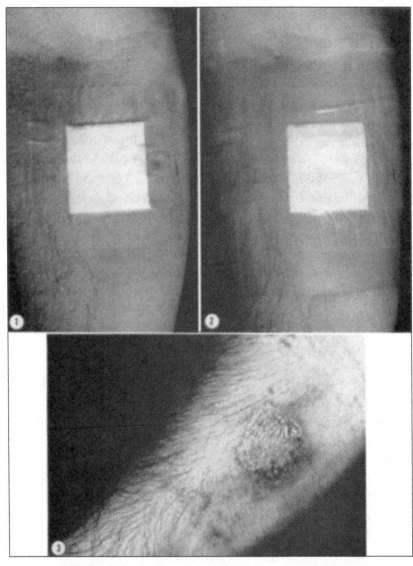

Figure 2.1. Maximization test on forearm. Copyright 1966 The Williams and Wilkins Co. Published by Elsevier Inc.

results, Kligman (1966a) vouched that "many thousands" of prisoners had participated over a six-year period, 90 percent of whom were black. This subject pool reflected not only the demographics of Holmesburg Prison but also the extensive processing and transferring of captives to and from a facility already above capacity at roughly 1,200.

As their designation suggests, "mild" or "weak" allergens are relatively harmless for most people who come in contact with them, but Kligman was interested in how these agents may harm those with more sensitive skin. However, rather than recruiting test subjects with sensitive skin, those he termed "hyper-reactors," he instead generated experimental settings purportedly replicating the condition. He achieved this by enhancing the injurious effects of normally benign substances, making them a lot more harmful than they ordinarily would be. What differentiated this maximization test from the usual patch test is suggested in its name, a process whereby the visibility of allergic reactions is amplified via the excessive amounts of test substances applied to the skin. Kligman called these settings "exaggerated," because they entailed creating extreme conditions of exposure. Laying beneath each square patch was a simulation of sensitivity that more or less guaranteed the production of injury. The experimental procedures leading to the maximization method did describe persistent physical and chemical assaults to the body's surfaces, typifying the planned production and distribution of pain common to experimentation: a "prolonged, *magnified* exposure of *deliberately damaged*, and therefore more permeable, skin to high concentrations of the test substance" (Kligman, 1966b, p. 404, emphasis added).

So to develop his new test, Kligman first intensified the destructive capabilities of mild chemical substances by boosting the concentrations at which they were normally found—in his experiments, the application of weak allergens bore "no resemblance to normal use" (ibid.). Second, he turned to modifying the skin itself, "deliberately damaging" it so that it was more likely to produce allergic reactions. For this he investigated several methods, including irradiating the skin with ultraviolet light, creating and cutting off blisters, flash-freezing the skin for ten to thirty seconds, and repeatedly using scotch tape to strip the skin of its outer layers. Finally, he arrived at using sodium lauryl sulfate (SLS), a surfactant and emulsifier present in the majority of cosmetic products and industrial cleaners sold today. Kligman weakened the skin with SLS by applying amounts far, far higher than what consumers were normally exposed to, resulting in skin with less protective capacities that was hence more vulnerable to the elevated doses of weak allergens later applied. Calling this process the "induction" period, Kligman essentially mechanized the irritation process to help augment the adverse potential of his test agents: step one, chafe the skin with SLS; step two, continue the onslaught with higher concentrations of allergens. This made resulting skin injuries more clearly obvious to the naked eye, to the light microscope, and to the photographic lens, with the observer better able to perceive the hazardous effects of weak allergens not through any improvement in visualization techniques or in the creation of new instruments of seeing but through the purposeful distortion and dramatization of the object of study—pain. Pain's heightened expression evidenced the skin's enhanced abilities to detect and respond to new agents, pushing to the limits its representational efficacy, or what might be called its immanent vitality.

The complete induction procedure, which Kligman called a "bombardment" of the skin, entailed alternating treatments of SLS and the test agent. For each patch-test site, five twenty-four-hour exposures to SLS were intermittently replaced by five forty-eight-hour exposures to the test agent, again, both at magnitudes much greater than normal use. This cycling between irritant and allergen resulted in tissue "crowded with lymphocytes [a kind of white blood cell] and dilated vessels engorged with blood" (Kligman, 1966a, p. 383). Sometimes, SLS was omitted when the skin became "too inflamed," exhibiting "vesiculation, oozing or erosions," or when the test agent was by itself already able to physically impair the skin (1966b, p. 396). Through this rotating contact with SLS and test agents, the skin barrier gradually lost its integrity and allowed test agents to more easily encounter immune cells underneath, priming or sensitizing the latter to their every appearance thereafter. By the fourth or fifth allergen patch in this induction process, a "sudden exacerbation or flare" would indicate a sufficiently sensitized immune system now able to mount a defense against the test agent—an allergic or inflammatory response.[8] The last photograph in Kligman's display of the maximization test shows the successful completion of this induction period, with the patch on the forearm now removed and revealing a swollen, blistered, and seeping lesion that Kligman classified as "moderately" inflamed skin.

The skin was allowed to recover following the induction period, but after recovery, it was subjected to the final step of the maximization method, a "challenge" or "provocation" test. This meant reexposing the healing skin to the same test agent, though this time, the concentration of SLS was even higher—doubled—since reexposure would occur only once and using more SLS would ensure that the skin barrier was adequately compromised to let the test agent sufficiently penetrate it. After forty-eight hours of challenge, the patches were read and the results scored and compared to the known allergenicity of other substances. Reading the injuries resulting from the maximization method relied primarily on visual interpretation, with low and high sensitization potentials, or "weak" and "extreme" allergens, identified through a five-point grading system based on the severity and frequency of observed skin trauma. Reddening due to heightened blood flow to the area signified a "minimum" positive allergic reaction, while strong responses included abscesses and swelling extending beyond the test site.[9] The greater the number of test subjects sensitized to the agent, the higher its allergic potential.

Screening Pain

Evaluating wounds that had been produced through the maximization test relied predominantly on the expert's eye, often generating inconsistent readings and thus forming a significant challenge to the accuracy and reliability of visual analysis. For example, observers often misconstrued irritation with allergic responses because they looked very similar. The recurrence of false positive readings—reactions interpreted as allergic when they were not—demonstrated the rather subjective nature of a scoring system whose "constant vigil" over different kinds of skin trauma was undermined by individual mishaps (Kligman & Epstein, 1975, p. 235). Researchers performed biopsies and histological analyses to distinguish between irritation and allergy, though here as well, the frequently "borderline" differences between them frustrated visual readings.

Researchers also contended with multiple variables influencing test results, such as infection, humidity and temperature, and duration of exposures. However, among all possible sources of uncertainty, the role of racial difference drew the most attention, emphasized in the design and interpretation of analysis. Repeatedly noting that the test and its results were attained using black skin—"Caucasians comprised not more than 10% of the volunteers" (1966a, p. 386)—researchers claimed that race significantly impacted the test's performance, or its capacity to visualize injury. "It is widely believed that Negro skin is more resistant to chemical attack," Kligman noted, "but the proof of this is not complete" (ibid.). He further asserted, "Pigmented skin, of course, obscures erythema [reddening] and may enable marginal sensitizations to go unnoticed" (ibid.). Black skin conceivably skewed test results on at least two accounts, the first being its assumed exceptional hardiness and the second being its color, which muted a visual indicator of harm biased toward less pigmented skin: redness. Impacting the outcome of experimentation, race was a complicating factor comparable to other experimental choices such as determining where, what, and for how long a test agent was to be administered. Like these other elements of study, racial difference then became a question for controlled investigation and analyzed for its potential to lessen the sensitivity and accuracy of the maximization method.

Worried that an apparatus developed using black skin would be ineffective when deployed on white patients, Kligman tested new protocols aiming to correct for racial differences by comparing rates of sensitization between black and white test subjects. As with his tretinoin experiments, Kligman once more excluded "very light-skinned" black prisoners, whose complexion complicated his aim to generate mutually exclusive or "statistically significant" results between races. Pigmentation again played a significant role in the experimental setup whose working definition of race was therefore based on color contrast, a categorical separation between light and dark skin that made no room for shades in between. An object directly produced through intentional choices, emerging with rather than before classifications of skin color, race in the maximization studies was inadvertently revealed to be on rather shaky ontological foundations, its borders made visible and porous by the simple addition of more skins. Yet its lack of material foundation did not render race moot, because it remained a concept or a tool that performed the boundary-making work at the core of Kligman's questions, the relationship between phenotype and the visibility of pain.[10]

In these studies, Kligman found that "the average intensity of the allergic patch reaction to the same allergen is clearly greater in Caucasian skin; intensely inflamed vesicular, spreading lesions are often produced in the Caucasian by allergens which induce only mild to moderate reactions in the Negro" (1966a, p. 387). Compared to black skin, white skin appeared to be more vulnerable to weak allergens, more prone to their harmful effects. But as with his tretinoin studies, Kligman again underscored the role of skin color in the presentation and visual interpretation of such results, which did *not* point to biological differences between white and black skin; this was not Kligman's concern, and so the proof of it remained incomplete in his work. The results instead demonstrated different capacities between white and black skin to *show* the kinds of reactions researchers were looking for: "The lower sensitizability of the Negro is probably *more apparent than real*. The difference is not immunologic, but one of lesser reactivity to an inflammatory stimulus. . . . By biopsy study, clinically marginal reactions are quantitatively the same in Negroes and whites" (pp. 386–387, emphasis added). At the cellular level, mild reactions to test agents varied little between white and black test subjects, as demonstrated in histological analyses that showed such reactions to be "quantitatively the same" in both groups. Under the microscope, white and black skin materialized inflammation in similar ways. The influence of racial differences in the formal classification of weak allergens therefore relied on skin directly accessible to the naked eye, which Kligman had elsewhere described as "that incomparable research instrument."[11]

Kligman later departed from this esteem he held for the eye, elaborating in his 1991 treatise on "invisible dermatology" an approach that countered the then prevailing assumption among practitioners and researchers that skin was like a window or a mirror to its own workings, unable to hide physiological secrets and so requiring no "high technology" to root them out. But now calling skin a "leather jacket," an opaque object covering up more than it reveals, Kligman determined that the naked eye had become an "unreliable instrument for judging the normalcy of skin"; histology and microscopy became Kligman's specialized tools of visualization, whose adoption would garner him the reputation of having transformed his field into an investigative science, with one colleague championing his methods as leaving "Blind Man" dermatology

behind.[12] Also emphasizing the centrality of vision to knowledge production and echoing this devaluation of the eye as unseeing compared
to visualization technologies, Kligman described the window/mirror
image of skin as a mythology, "invariably antagonistic to the acquisi-
tion of sound knowledge," a misperception that had "kept dermatology
in a backward, disparaged state for at least a century."[13] Kligman had
already been making widespread use of histology and microscopy in
his experiments at Holmesburg Prison three decades before his writings
on invisible dermatoses, yet he minimized their value in maximization
tests, which he designed with and for the naked eye. Lacking in photomicrographs, Kligman's published images of the maximization tests
pin visual evidence solely on macroscopic phenomena. And given his
deliberate avoidance of black test subjects with lighter complexions, the
images also reflect his preference for using white skin when photograph-
ing injury. The forearm on display most likely belonged to a white test
subject, their wound thus coming to represent an experiment whose
substantial majority of test subjects were not white but black. Again,
these results did not conclude that black and white test subjects had essentially distinct immune systems; "The difference," Kligman reported,
"is not immunologic." More apparent than real, they instead followed
from the skin's ability to *convey* an immune response—"reactivity to an
inflammatory stimulus."[14] White skin, simply put, was better at showing
that it hurt.

In a separate publication evaluating quantitative approaches to measuring irritation, scientists more explicitly connected race to the durability of skin, stating that "Negro skin is more tolerant of chemical
irritants" and instructing that the experimental observer also "correct
for the *obscuring effect* of pigmentation" (Kligman & Wooding, 1967, p.
80, emphasis added). These experiments deployed the same assemblage
developed for the maximization method, as shown by a single photograph of one patch copiously taped onto the lower back (figure 2.2).
Unlike those in the maximization tests, this photograph of the irritation study is not accompanied by others revealing the harm caused by
test agents. The reader never sees what happened to the skin hidden
beneath the patch, left only with a singular pictorial representation of
procedure but not of result. As the publication assures us, however, the
experimental observer exercised a more "practiced eye" for detecting

Figure 2.2. Patch-test apparatus on the lower back for irritation test. Copyright 1967 The Williams and Wilkins Co. Published by Elsevier Inc.

the "purplish hues" that would signal local irritation in black skin, their observation, too, aided by the same intensification of effects adopted for the maximization method: increased amounts of test agents applied to the skin (ibid). Together, the observer and the experimental design coaxed reactions from an object deemed especially uncommunicative— black skin—perceived as undermining the recognition and therefore the standardization of harm. To scientists, it was physically stronger, "less responsive to exogenous insults," and it was secretive, "less capable of expressing the inflammatory change," masking tissue flushed and made tender by chemical and physical abrasions (1966a, p. 386). Black pain, in short, was simply too hard to see, confounding the acknowledgment of pain as such. Believed to be both difficult to wound and unforthcoming of evidence of damage, black skin was then subject to even more aggressive methods that ostensibly sharpened the visual resolution of pain.

Close-up photographs of the maximization and irritation tests spotlight the physical construction of experiments, images of white square fabric secured to skin with layers and layers of tape, constituting a mundane presentation and visual instruction of method. The pedagogical value of these photographs is their illustration of new apparatuses, new means of administering, modulating, reading, and evaluating the deleterious effects of allergens and irritants in direct contact with skin. Perturbed by the supposed ephemeral and deceptive nature of black pain, however, the photographs stage the ambivalent meanings of black skin, which was both an obstacle to interpretive accuracy and an important resource for making patch tests more efficacious. With black skin, pain caused by experimental procedures was claimed to be imperceptible, necessitating more "careful examination," even as what *is* seen, what the photographs are supposed to make readily apparent, is the procedural application of pain. The photograph of the irritation test, for example, is a straightforward image of this harm in process, a persistent and concentrated onslaught, and yet its accompanying text invites the viewer to *not see* what they are being shown, undercutting the photograph's didactic purpose and the efficacy of the standardized harm protocol it presents to the viewer—the contraption of cloth patches, plastic tape, and chemical agents robbed of its indexical power by the very target of its focused and intensified assault. Black skin damaged the representational power of pain.

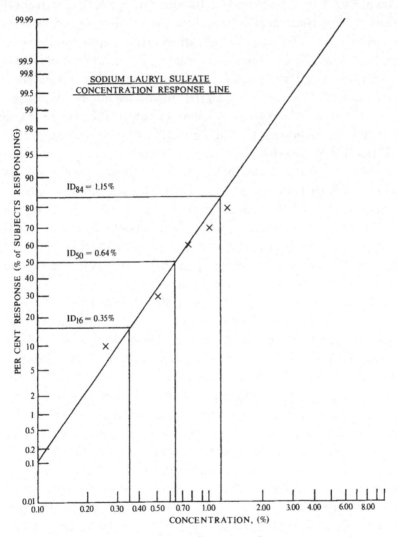

Figure 2.3. Graph showing concentrations of SLS required to cause irritation in 16 percent, 50 percent, and 84 percent of test subjects. Copyright 1967 The Williams and Wilkins Co. Published by Elsevier Inc.

More than photographs, Kligman's publications on the maximization and irritation studies teem with graphic and numerical evidence, representing in mathematical form results garnered from the new bioassays shown pictorially. Annemarie Carusi and Aud Sissel Hoel (2014) call the interplay between qualitative and quantitative imaging processes the "hybrid vision" of laboratory work, where photographs and measurements are mutually constitutive of meaning. Kligman's irritation study, for example, includes graphs illustrating the relatively linear relationship between the quality or value of an irritant and the number of test subjects who exhibited irritation. This value can reflect the amount of time the irritant remained on the skin. Or it can reflect the concentration at which it was applied (figure 2.3). In both instances, raising the value—increasing the time or concentration—led to greater numbers of test subjects with irritation. Kligman's graph of SLS plots these results and draws a straight line through them, sloping upward from left to right to show a correlation or a positive relationship between the amount of SLS used and the percentage of test subjects who reacted to it. Furthermore, the slope is bisected three times by perpendicular lines indicating what specific amounts of SLS produced a specific percentage of reactions, drawing the eye to more specific relationships and making them therefore more scientifically relevant than others that the graph also shows. This relevance is named directly, borrowed from the field of toxicology and applied here to skin experiments, the values $ID16$, $ID50$, and $ID84$, or concentrations of SLS that produce reactions in test subjects. As the graph indicates, 16 percent became irritated at 0.35 percent SLS, 50 percent at 0.65 percent, and 84 percent at 1.15 percent. Calculations like these, Carusi and Hoel argue, constitute a form of seeing, but they are imagistic not simply because they share with text the form of inscription but because they mediate through a different, non-pictorial register other than perceptual relationships contained in scientific looking. Here, the pain reported as optically inconspicuous and thus photographically absent is nonetheless countable and diagrammable, apprehended through figures drawn and tabulated in charts, tables, and graphs.

If the photograph shows method, the numbers show results, supplementing the former with a level of detail and precision afforded by statistical analysis; for example, 50 percent of subjects reacting to a 0.65 percent SLS solution. Transforming the singular object of the picture into a multitude, these graphics of pain convey their object through greater and greater abstractions such that pain is visualized not as wound but as information, harm transposed or reincorporated into a body of data. This is not to say that images of irritated skin would undo these mathematical translations or that photographs are somehow less lacking in abstraction than are tables and graphs. Epistemologically different from but still read alongside the pictorial, numerical representations of injury proliferate the photograph's work of abstraction. The pain simultaneously shown and denied in the image becomes tally, *undifferentiated by race.* Charts and graphs of the maximization tests do not include race even as race figures prominently in the text as a complicating factor. Missing from tabulations, race does not perform the invisibleizing work there that it does elsewhere in the scientific papers, as if its absence in data visualization is what allows pain to be counted. Hybrid vision in these studies joined a coordinated structure of laboratory vision extended like skin across bodies, instruments, photographs, and formulas along which the recognition of black pain was continuously displaced.

Kligman deployed the maximization construction to test hundreds of compounds over a period of fifteen years, after which he resolved to upgrade it specifically for white patients and white test subjects. Created using mainly black skin and tested on a mostly black captive population, the first design was deemed too caustic for white skin.

> The fact is the Caucasoid will develop painful, *intolerably* severe dermatitis from a relatively low concentration of [SLS] while the deeply pigmented Negroid may show no more than a little redness. . . . We often find lower reactivity even when the barrier is artificially breached by scarification [scratching with a needle]. The issue is not settled in our minds. Perhaps blacks are endowed with skin that is better able to cope with external physical and chemical insults. (Kligman & Epstein, 1975, p. 232).

Though remaining unable to explain the mechanism behind it, researchers held fast to the notion that white skin was more sensitive or more fragile and thus necessitated modifications to the maximization test. These included decreasing the number of times that test subjects were exposed to SLS during the induction period, cutting it down from five exposures to just one or two. White test subjects were also allowed twenty-four-hour breaks between each round of irritation in the induction period, allowing their skin to mend between SLS applications. The last modification pertained to the challenge portion of the test, where the strength of SLS was diminished by up to half of the original amount used for black test subjects. Altogether, these alterations meant that white skin would be subjected to both weaker and shorter bouts of chemical harm. The most significant change, however, was in Kligman's suggestion that white skin rather than black skin be henceforth regularly used for the maximization test, noting, "With potent allergens it makes little difference whether one uses blacks or whites; however, the latter are preferable for identifying weak allergens. For one thing, erythema is more *easily perceived*. Borderline sensitization may therefore be *missed* in the black" (ibid.). While only ever speculating on the supposed toughness of black skin, Kligman was unequivocal about its inadequacies as a visual instrument. It concealed or made inscrutable the phenomenon that white skin clarified, though that did not stop researchers from harnessing it for purposes now later admitted to serve whites. The privileging of whiteness as a cipher of pain, Rebecca Wanzo (2015) shows, conditions iconographies of suffering shaping the visibility of black injury.

However, though white skin was a more desirable medium for displaying inflammation, it also introduced its own interpretive challenges, mainly that its wounds seemed to overwhelm the expert eye with too much information. Whereas the difficulty of registering inflammation on black skin meant that it could be "missed" or overlooked, the gross intelligibility of injury on white skin meant higher incidences of false positives or of "non-specific reactions." "We have a firm impression," the researchers attested, "that the variability in the range of responses to irritants is greater in whites than in blacks" (ibid.), with white skin exhibiting injuries not only qualitatively different from those seen in black skin—they were "intolerably severe"—but also more varied or numerous in type—they were diverse. White pain, in short, was simply too "easy" to see, so plainly obvious as to require a parsing out of the assortment of injuries it could manifest. Exceeding Kligman's visual indicators of inflammation and thereby cluttering his data, white pain and its deleterious effects on experimental accuracy called for mitigation, hence the modifications to the maximization test that collectively softened its chemical attacks. These adjustments demonstrated not only the inverse relationship made between pain and race but also the permissible scope of intervention. While the detection of black injury meant disciplining the expert's eye, making it more sensitive to or adept at catching something presumed transient, the relative ease of seeing white pain called for altering the research design itself, supplanting individual training for structural changes that established a level of comfort taken for granted in black test subjects. Enhancing the visualization of "mild, borderline" reactions against black skin figured as armor, the maximization test and the visual metrics of harm it sought were thus predicated on mitigating white pain alongside hurting black bodies even more.

Captive Agency

Kligman had maintained the relatively safe development of his first maximization method, which, we are told, had been tolerated "with little complaint" by thousands of black test subjects and in whom no serious or long-lasting impairments had been ascertained. He claimed only one exception: the accidental administering of a dangerously high concentration of a test agent that then caused "tachycardia [increased heart rate], headache, nausea, vertigo, and altered emotions" in test subjects (1966b, p. 404). Unsurprisingly, Kligman's articles do not report conflicts occurring between researchers, prisoners, and prison staff, instead assuming a language of collegiality portraying Holmesburg as an environment of collaboration among all parties involved. Acknowledgments say, "We are indebted to the inmates of Holmesburg prison for serving as volunteers and to the administration [Edward Hendrick, superintendent] for use of the facilities."

In the same year that Kligman published his findings on the maximization test, Holmesburg Prison invited journalists to witness and cover the large research program that had now operated for close to a decade behind the prison's walls. The coverage provided by the city's top newspaper, the *Philadelphia Bulletin*, echoed the spirit of comradeship suggested in the acknowledgment sections of Kligman's publications, as seen in its front-page photograph providing a snapshot of a medical examination taking place in the prison's H Block, the corridor constituting

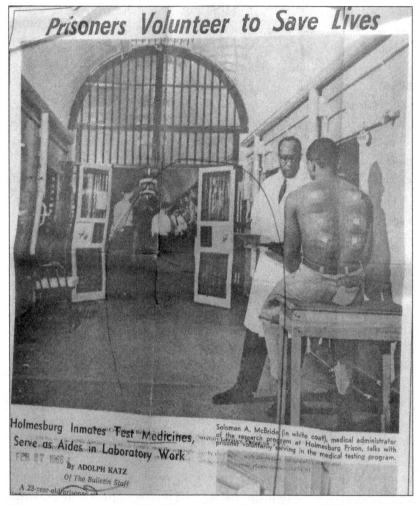

Figure 2.4. Front-page news photograph of medical examination in H Block. Created by Michael Maicher for the *Philadelphia Bulletin*. Courtesy of the Special Collections Research Center, Temple University Libraries, Philadelphia, PA.

the main operating hub of Kligman's research program (figure 2.4). At the center of the photograph are human and architectural forms of spatial control—a prison guard keeping watch at one of the several metal gates partitioning the long passageway. Closer to the foreground, several feet ahead of the gate and its sentry, are two figures, one named in the news article and the other not. The latter is a shirtless prisoner sitting, slightly hunched, on a wooden bench or table with his naked back turned toward the viewer. He faces the experimental program's medical administrator, Solomon McBride, who, in contrast to the prisoner's state of undress, is fully clothed and draped in a long white lab coat. Through his dark-rimmed glasses, McBride gazes down at a clipboard he carries while he stands speaking to the prisoner, quite possibly discussing the six white gauze patches lining the latter's back, three on each side—patch tests, presented in long shot for the lay viewer. Placed right above the image, the article's title invites this viewer to read in it the good will and selflessness of those behind bars: "Prisoners Volunteer to Save Lives," medical progress made possible through captive humanitarianism.

The news story does not elaborate on the tests that the prisoner in the photograph is undergoing, instead enumerating other tests in which approximately 900 of Holmesburg's 1,200 captives were taking part:

> The man in isolation helping to find a cure for jungle rot has voluntarily put himself "in the hole". . . . [He] may remain there for a week or ten days under close supervision, fighting boredom and monotony as the medication permeates his skin. . . . In another cell, men soak their hands half an hour each in a solution intended to test the basic ingredients of toothpaste and washing powders. . . . Other prisoners sit under "black light," subjecting themselves to photo-sensitivity in tests to determine the effects of sunlight. . . . You walk through other cell blocks . . . and you reach another room where prisoners are lined up to give samples of their blood. (Katz, 1966)

Interpolating the reader as a visitor to H Block, ambling from cell to cell and inspecting the happenings in each one, the story shows the prison hall to be abuzz with activity, presumably facilitated in part by the prisoners themselves. They "voluntarily" agree to solitary confinement and queue up to submit their bodies for experiments, while others "serve as aides" through recordkeeping and assistance in enrolling and administering tests to their counterparts. Kligman promulgated the rehabilitative dimensions of these activities, stating, "Very often, the more serious a crime a prisoner has committed the more willing an individual he becomes in our work," suggesting the prisoner's desire to absolve his guilt or repay the "debt [he] owes to society" by participating in experiments that may go on to help others. A few who did not become test subjects still reaped the proclaimed remedial benefits of Kligman's experiments by taking up supportive roles. Of the nine hundred prisoners involved in Kligman's research program at the time, roughly one hundred served as technicians whose many duties the article enumerates: "check blood pressure and take blood samples," "assisting doctors in routine examinations," "typing case histories," "take dictation in long hand," "take skin scrapings and perform biopsies," "assists in writing up reports," "keep an inventory control on the stock," "helps spin down the blood samples and extract plasma," and so on. For researchers, these tasks performed by prisoners also doubled as "teaching," establishing the pedagogical role of Kligman's program. The prisoner as technician can, too, learn how to carry out lab work. "He may be ragamuffin on the outside," said Kligman, "but in two months we can make a highly skilled technician out of him."

From these descriptions, Holmesburg Prison is barely recognizable as a prison, spaces known for their intense surveillance, deprivation, and physical and spiritual violence. Like Kligman's publications, they evade or cover up these tensions inherent to prison existence, replacing them with an image of captive agency showing prisoners to be active and willing participants in the conduct of experimentation. Popular or scientific, they constitute a mediascape conducive to penal spectatorship or a cultural understanding of captive pain that Michelle Brown (2013) argues distances that pain from public life even as it enters the public arena in its mediated forms. Still, the article's narrative fails to cover up or foreclose all of these tensions, irrupting with the same ones also permeating Kligman's maximization experiments. In the latter, black skin was concomitantly capacious and disadvantageous to the aims of research, a contradiction that nevertheless did not impede or lead to the cessation of study, because it catalyzed an aggressive reworking of black skin to the demands of research, forcefully adapting it to visual signs of pain exclusively defined by observers. One can see this same conceptual maneuvering at work in the article's double-sided portrayals of prisoners, who go unnamed but are pointed out one by one as the reader vicariously tours the research program. The prisoner over there, carefully checking blood pressures and taking blood samples—he was in "for aggravated robbery involving a taxicab driver." Another assisting doctors had "a long history of drug addiction" and had also assaulted a police officer. And the one typing up case histories? He committed forgery. The one meticulously distilling plasma from blood? A burglar. McBride's personal favorite technician? A killer. (The article speaks with a certainty that elides the *un*certainty of legal culpability that marked most prisoners at Holmesburg, a significant majority of whom were charged but not convicted.) Veering between descriptions of crime and descriptions of competency, the article generates a whiplash of meanings hardly imparting a moral lesson about the redemption of those locked away from society, though the article tells a bit of that, too. Taking every chance to remind the reader that "Holmesburg volunteers [were] criminals of all types," the article ultimately reciprocates the same principle underlying Kligman's maximization tests, that is, the reconfigured usability of problematic human material. The news article does not let us forget that these capable technicians were nonetheless inveterate wrongdoers—they were "armed robbers, forgers, killers, strong-arm men, thieves." Just as black skin remained useful even as it ostensibly compromised that which it was useful for, the abilities and productivity of Holmesburg's prisoners did not transform or erase their deviant status but rebounded on it.

This symbolic relay in the article's text finds its visual complement in the front-page photograph, where the eye travels back and forth between the looming metal gate at the center of the frame, and the dyad of researcher and test subject off to the side but in front of the gate, situated much closer to the viewer. This composition diminishes the background and figuratively may not even have one, with the prison architecture as prominent a subject as are the researcher and the prisoner. Though they take up different parts of the photographic setting, the gate and the medical examination taking place before it assume equal visual weight, creating an unbroken and reversible line of sight between a scene of science in action and the built environment of incarceration. Through both text and image, the article repeatedly informs who and where potentially life-saving knowledge was coming from, but this story is not about atonement or rehabilitation. It is about the irredeemable criminality that produced expedient lives for scientific and medical progress. Indeed, with his back turned toward us, the prisoner may not be a visual subject in his own right but a supplement to the figure of McBride, becoming an extension of or an accessory to the researcher's activity. Composing the background to six white patches that figure more clearly than he does, the prisoner's naked and wounded back is not so much an object to be looked at but an imaginative expanse for seeing one's place in biomedical progress, picturing advances to which the reader stands as beneficiary.

This interplay of oppositional meanings that followed images of cloth patches on black skin came to signal the biological and social dimensions of captive agency, chimeric in its expression. Inexplicably more sturdy, black skin made poor experimental material for the new and successful bioassay it developed anyway, being both generative and resistant to this generativity. Captive participation in such experiments were at once indicative of self-sacrifice and perhaps of attempts at absolution, but also signifying or transmuting an intractable degeneracy. Kligman psychologized his technicians and test subjects, saying:

> Many of the prisoners, for the first time in their lives, find themselves in a role of important human beings. We say to them, "You're important, we need you." Once this is established, these guys will knock their brains out to please you. If the experiment does not pan out, they get depressed. They become emotionally involved in the project. . . . When a criminal does a bad job, he feels bad. He expects you to mistrust him and is astounded when you believe him. He becomes like a monk, a super-purist, and runs to you like a kid displaying a scrupulousness you just don't believe. (1966b)

The diligent participant was nevertheless infantile, his proficiency in lab work or faithful compliance to procedures stemming from a childlike earnestness to please authority rather than from training and accumulated competency. A monk or a super-purist, he reproduced the culture of ascetism emblematic of laboratory work, but this disciplined attitude only channeled his juvenile predisposition, ready to knock brains out to satisfy supervisors who must then act as paternal figures. Saidiya Hartman (1997) reveals how such accounts of captive agency reinscribe the domination and brutalization of captive subjects, their insidious deployment of volition bounding captives ever tighter to authority by attributing the captive's submission to self-motivated forms of enjoyment and self-making. This is not, in the context of Kligman's research program, to deny the actual presence or the lived experience of captive pain and agency but to see both popular and scientific images of patch exams as fraught expressions of willed subjection, what Hartman terms "ruses of power." Redoubling captive subjection, positive representations of agency in both human and nonhuman instruments, or affirmative gestures toward the agency of prisoners and of the skin apparatus were as revealing of racialized contexts of exploitation and domination as were representations demonizing that agency. Connecting the expert's gaze with that of the lay viewer, agency in its human and instrumental appearances and in its positive and negative guises meant the absence of captive suffering and by extension the impossibility of abuse, giving scientific and popular images of patch exams their form as visual artifacts within a scopic regime that, to borrow from Elizabeth Abel (2014), "denies [its] subject the right to be irritated" (p. 114).[15]

Consider the following statement by the director of the National Institute on Drug Abuse Addiction Center during the 1970s: "The prisoner patient is often an extremely egocentric individual who is much concerned with his self-interest and is quite able to reject any experiments that he does not see to his advantage."[16] The prisoner's unique psychological profile, he warned, was linked to forms of "psychopathy"— alcoholism, drug abuse, criminality—making them unable to "meet many of their responsibilities to society, and in many instances their participation in research is one of their more constructive contributions." Hence a conduit for redirecting antisocial impulses, research programs also introduced new modes of prison management, helping to pacify and control the prison population by channeling their self-centered interests into socially condoned behavior. Couched in less technical language, the pathological root of captive agency was also alluded to by one prison administrator, who described such assertions of captive will as acts of "conning."[17] Kenneth Hardy, then head of the Washington, DC, Department of Corrections, expressed incredulity that prisoners had joined experiments out of shame or charitable feeling, insisting instead that at least some of them were "shucking and jiving" to maintain their place inside a lucrative prison industry. Recalling nineteenth-century blackface minstrelsy and, before that, the stereotype of the happy slave, Hardy's portrait of captive test subjects inverted relations of domination where the scientific management of prisoners was at the same time a captive performance of deception against authority. Captive test subjects were said to be putting up a front, an act, a disingenuous expression of humanitarian spirit that did not merit concerns of abuse precisely because it made them particularly suited to scientific study. The reference to minstrelsy is telling given that such imagery, Michelle Ann Stephens (2014) shows, became the epidermal sign or framework for reading colonial logics onto the black body. Compared to caricature or reduced to black stereotype, captive expressions of agency in prison experiments complicated the recognition of authentic feeling.

So even when prisoners did adopt the rehabilitative and redemptive discourse of prisons and research programs, their actions and reasons for them were seen as inauthentic goodwill, turning the stated institutional benefits of imprisonment and experimentation into vehicles for captive trickery and manipulation. Moreover, as Anthony Hatch's (2019) sociological study of prison experiments show, the ascribed altruism and patriotism of prisoners often fell along racial lines, more validated in white test subjects whose participation might have, in fact, enabled efforts to construct testing programs as captive demonstrations of generosity and citizenship. Coded white, these programs could be seen as "strengthen[ing] the nation by speeding up technoscientific progress" (p. 74).[18] The skin color of test subjects determined the believability of captive claims of public service. But where whiteness signaled breakthrough and development, blackness threatened and appropriated for itself the possibilities of technoscientific enterprise, thus, to borrow from Stephens, "recording symbolic understandings and codings of blackness as, primarily, *a visual sign of the difference of the racial other*" (p. 38, emphasis added). From Kligman's scientific images to broader narratives about prison research programs, the *appearance* of captive agency (its mediated forms) were heterogenous and conflicting but sutured together by the unthinkable or unseeable condition of black pain against the ineluctable vibrancy of white suffering.

"Acres of Skin"

Bracketing out relations of enmity between captives and their carceral and medical overseers, depictions of voluntary participation among

Holmesburg's prisoners were in marked contrast to the pervasiveness of experimental sabotage that had occurred among them. From his interviews with former Holmesburg guards and prisoners, Allen Hornblum gleaned many instances of "cheating" Kligman's patch tests.

> Guards said it was not unusual to go into an inmate's cell and find long strips of adhesive tape hanging from the wall. Thomas Sims, 18 years old when he first entered Holmesburg in 1966, confirms that "some guys took [the grid of patches] off and hung it on the wall. . . . Those doctors were running a game on us, so we ran a game on them." Tom McGevren, a former lifer who entered Holmesburg in 1972, concurs: "To some degree it was a joke because we wore the patches in the testing block [but] took them off and hung them when we got back to our cell." (1998, p. 12)

Among test subjects for whom the skin off their backs literally comprised their sole object of exchange, "games" of deception helped navigate the program's formal economy and forms of exploitation it introduced to the prison, ran around on untrusted researchers busy capitalizing on a caged population whose certain and absolute submission they had either mistakenly assumed or inflated for the program's public and professional image. The ramifications for the medical knowledge produced were not insignificant. On the contrary, acts of subterfuge like peeling off patches or unexpectedly withdrawing from experiments thwarted the orderliness of experiments and the accuracy of data collected, essentially transforming research procedures into "jokes" or contests of duplicity that led one prison reformist to call some of Kligman's studies "scientifically worthless."

Such transgressions in protocol may have contributed to Kligman's findings concerning racial differences in skin sensitivity, with the unwillingness of test subjects to keep their patches on consequently diminishing their exposure to harm. Yet test subjects did not remove their patches for mere mischief or for extracting some modicum of amusement or retribution from scientists. They stripped off their patches because they were in agony. Again, from Hornblum's interviews:

> "That first test nearly killed me, it was so painful. I nearly went through the wall. I had a patch put on my back that covered a large area. It was

a 10-day test and I wasn't allowed to take a shower". . . . "I looked like a checkerboard with patches and skin discoloration on my arms, back, and chest". . . . "I was scared of the patch test, it was like a tattoo. . . . One year later I got on one [study] and I was sorry I did" . . . "blisters made it very uncomfortable to lie on [my] back". . . . (pp. 10–26).

Receiving little care from researchers, test subjects were left to bear or to relieve their own pain, the same burns, sores, and abrasions disavowed by Kligman and his team even when such injuries came to permanently mar the skin. If data from his maximization method were erroneous, then it was likely more a problem of apathy and negligence on the part of researchers than of their test subjects' defiance of procedures and aversion to taking on more harm. Skewing results or derailing experiments entirely, these local forms of subverting the goals and methods of research challenge interpretations about prisoners passively accepting direction from scientists. Ultimately, the nature of weak allergens and irritants was not to be revealed by expert vision or in recorded evidence and scientific interpretation. Rather, it issued directly from test subjects themselves, these non-experts whose felt experiences of pain went unrecognized as demonstrable knowledge even as that knowledge was directly borne by the wounds and scars that checkered their limbs and backs. Branded skin was the injurious constitution of captive agency or, to borrow from Alexander Weheliye (2014), a corporeal de/formation of agency in the absence of freedom.[19]

This agency discussed in the chapter, an agency that is fundamentally imagistic, encompassed both the photograph and the skin apparatus, skin appropriated from its body and transformed into another medium of documentation. Building on Michel Foucault's notion of "capillary power," which displaces power from centralized protocols to its operations in and through bodies, Dylan Rodríguez (2007) locates the capillary power of carceral violence in the imprisoned subjects' viscerality, their blood, skin, nervous system, and organs functioning as mediating materials of the prison regime. The skin apparatus literalized this capillary power, pointing to the enmeshment of biomedical practice with carceral strategies of containment, surveillance, and control. Forming the material inscriptions of capillary power, the captive body corporealized not only the doing of science *in* prison but also what Foucault ([1977]

1995) had discerned as the science of imprisonment itself; to wit, the classification, differentiation, supervision, and codification of criminal pathology.[20] And if, as Don Ihde (1983) also argued, instrumentation extends, embodies, and therefore ontologically precedes our experiencing and understanding of the world, then the captive body as research object and device became the means for defining, bringing forth, and mastering the nature of the penal system and the lifeworld of postwar human testing.

No longer the outer, protective envelope shielding the subject from external harm, skin taken from the captive body affirmed its representational authority by denying the pain of its removal, by becoming for scientists the perfect amalgam of object and instrument of the look. And is this skin even "skin" anymore when it has no body, no form, nothing for which it can act as skin? Tellingly, Kligman had said about his first visit to Holmesburg, "All I saw before me were acres of skin. . . . It was like a farmer seeing a fertile field for the first time." Among prisoners, skin failed to signal corporeal boundaries, instead forming the material basis of their de-individuation into expansive, usable terrain: "'an anthropoid' colony . . . which wasn't going anywhere." Skin is generally thought of as a sensual surround that, while containing the inside from the outside, also helps cohere a sense of self by mediating the body's interactions with the world.[21] Rather than demarcating individual personalities or enabling relations of reciprocity, however, the skin of Holmesburg's prisoners composed an undifferentiated, fleshy landscape, what David Marriott (2007) calls a "displaced-condensed" figure caught somewhere between symbolization and its failure. Captive skin was not a contact zone between self and other but an apparatus and a place where biomedicine and incarceration could touch and fold into one another, their shared routinization of pain and suffering dissipating into standard operating procedure.

3

The Skin of Architecture

Holmesburg Prison is haunted. I did not know this when I first arrived at the decommissioned facility to study first-hand its deteriorating architecture until one of the photographers on site joked, "Seen any ghosts yet?" On my second visit, I learned that yet another photographer, post-shoot, had discovered in several of his developed pictures a faint, phantom-like figure of a man looking directly into his camera. This story, however banal as ghost stories go, was a frightening start to a largely solitary exploration of Holmesburg's crumbling cells and corridors. Later, a quick internet search revealed that the spirits of Holmesburg were indeed restless, their "poltergeist activity" assuming "pretty much every story you can imagine, with tales of brightly colored orbs and strange sounds to more bold claims [about] apparitions of prisoners appearing to charge at people."[1] "Neighbors," one blogger notes, "have reported hearing loud screams . . . gunfire and riots" coming from behind the wall enclosure, and that most who come to the prison "get the feeling of a heavy energy throughout."[2] Inevitably, these stories also reference Albert Kligman's research program as a key source of ghostly ire, the prison's history of medical experiments now retold through spooks and thrills.

Holmesburg Prison is a dump. Empty soda cans, plastic water bottles, Styrofoam containers, paper cups, cigarette packs, film canisters, and more litter the prison grounds, testament to Holmesburg's status as an abject or forgotten place. City officials had promised residents of the surrounding neighborhood that Holmesburg would be demolished when it closed in 1996, and yet over two decades later, it still stands with walls decaying and shreds of building progressively crushed underfoot, this waning structural integrity and heightened risk of harmful chemical exposure restricting access to only those visitors willing to sign liability waivers.[3] The whole edifice is in a much greater state of dilapidation than Eastern State Penitentiary, which predates Holmesburg by nearly a century and after which Holmesburg and three hundred other prisons across the globe were modeled. But considered the world's first modern prison, Eastern State, unlike Holmesburg, is subject to ongoing restoration and conservation projects proper to national historic landmarks. But this added to Holmesburg's "cool" and "creepy" factor, which had convinced the director of the 2017 horror flick *Against the Night* to choose Holmesburg over Eastern State as the film's location. Narrating a ghost-hunting trip gone wrong, the movie needed an otherworldly setting that Eastern State could not provide. "Eastern State is awesome, but some of it is a little too much like a museum in places," he said. "It's kind of like Holmesburg's little brother, and Holmesburg is just straight-up abandoned" (as cited in Vadala, 2017).

So here was my archive: a haunted prison in ruins and refuse. I went there expecting to see and record a history of experimental abuse embedded in its weakening architecture, to witness how metal, stone, and concrete could objectify the intersections of captivity and medical science, or to observe how the built environment can manifest these entwined histories and logics. I was uncertain whether a hidden story was waiting to be unveiled from within Holmesburg's material form or if this story was readily apparent, splayed out on the surface of a building whose inside and outside are in many areas no longer distinguishable. In either case, I expected (or wanted?) Holmesburg to "speak," to show, or otherwise to materialize something about itself that it no longer is: a prison space and a laboratory site. This approach is further complicated by Holmesburg's new uses in memory work. It was the filming location of a number of music videos and five Hollywood motion pictures.[4] It remains the subject of many a photographer's online portfolio. It has inspired ghost stories and was even featured in a 2014 Destination America miniseries on paranormal activity. And in 2011, it was the site of a major art conservation project that sought to preserve the building's vanishing graffiti. As the current chapter will discuss, many of these present narratives about Holmesburg refer to Kligman's research program, repeatedly citing it and becoming a significant source through which Kligman's work becomes more popularly known. Notably, the art conservation project compared the skin of Kligman's experimental subjects to the run-down walls of the prison or, as the project description stated, to the "skin of architecture."[5]

Taking up this spatial notion of skin addresses the logical and topographical continuities between sites of captivity and sites of medical research, both of which can be understood as interlocking disciplinary and knowledge-making (eco)systems. In some sense, this more expansive notion of skin decenters the individual by spatializing the body, corporealizing the material landscape, and imagining an interiority inherent to space that is not reducible to the human subject. While this dovetails into current dialogue on the immanent vitality and agency of matter, the skin's movement from the captive body to the structures that caged it brings a more somber tenor to laudatory renderings of material life that often overlook subjects who have historically and ontologically occupied the position of things. As discussed in the previous chapter, life conceptualized in this new materialist turn to objects and matter may run counter to and even reify the subject who embodies the gap between "human" and "being," failing to recognize the figure already at the kernel of its most critical interventions in humanism. However, tracing the skin of Holmesburg's architecture does not promise to locate this subject-object, and by extension troubles any straightforward connection made between medical and prison geographies. Holmesburg is now neither of these, its history as prison space and laboratory site readable only through its material disappearance, resident hauntings, and steady accumulation of detritus. At Holmesburg, I went to see science and came away with ghosts, garbage, and ruin.

Ruin Photography

The modern surface, English scholar Anne A. Cheng (2011) writes, implicates a relationship between architecture and skin, a relationship that not only projects the notion of skin onto a building's surface but also invokes oppositions made between interiority and exteriority that are always deeply racial. Modernity's dream of a pure surface, a white surface, is inextricable from troubled visualizations of black skin, signaling a point where "aesthetic history meets the history of human bodies made inhuman" (p. 12). These interlocking histories are not, however, simply marked by repression or exploitation of blackness, for the multiple surfaces modernity assumes (architecture, clothing, the image) also mark the ways blackness comes to disturb the very terms of representation itself. Holmesburg's past and present uses implicate these complex histories, though, in this case, aesthetics is less an object than an archive, a collection of the varied ways Holmesburg has been documented and through which certain human subjects become known or remain unknown.

The most accessible archive of Holmesburg's prison space is that made available on independent websites by professional and amateur photographers interested in making ruinscapes, or imagery of abandoned structures in significant disrepair. Though not open to the public, Holmesburg hosts limited photo tours in the summer months, allowing photographers to record the prison's state of decline.[6] Such images are often panoramas and wide shots of hallways, cells, and staircases all in a slow process of collapse, revealing rusty beds and gates, delicate layers of peeling paint, exposed piping, clutters of debris, scraggly vines and roots twisting around and in between partitions and rooftops—each an arresting spectacle of the death of a building or of solid surfaces undone. Capturing the interior landscape of structures in all their captivating wreckage, ruinscapes have been pejoratively dubbed "ruin porn," a coinage first launched at pictures of Detroit's abandoned buildings, which number in the tens of thousands. Taken and popularized by reporters, artists, and photo-tourists in the last decade, these images arguably elide or even fetishize the city's economic blight and growing, disproportionately black poor, their "postapocalyptic feel" exploiting the motor city's downfall for the manufacture of awe and nostalgia at modernity's decay.[7]

There is surely something pornographic about visualizing disintegrating surfaces—surfaces penetrated and degraded by the elements—and made even more so by what is left unseen; that is, the very real human struggle and suffering definitive of structural marginality, the status of being socially left behind. In this sense, ruinscapes like those of Holmesburg replicate the structural erasures that come with abandoned places, which are made to stand in for and blot out the people connected with them.[8] But as undoubtedly with porn, the pleasure and titillation producing and produced by ruinscapes cannot be easily dismissed as problems of fancy, bad taste, or false consciousness. Rather, the pornographic and its attendant affects, too, constitute avenues for reaching the changing ideological histories of structures. Ruinscapes are inescapably problematic in their intention, to paraphrase one

Figure 3.1. Aerial photograph of Holmesburg Prison by Aero Service Corporation, 1922. The Library Company of Philadelphia.

blogger, toward "abandonment" and not "the abandoned."[9] However, they are also inescapably part of the representational economy of deserted places like Holmesburg, this intertextuality complicating what might be viewed as strictly positive or negative interpretations. Ruinscapes can hence be neither only celebrated nor only ignored, the dissolution of boundaries made most palpable by decomposing surfaces perhaps also signaling the indissoluble link between the problematic and the counter/alternative, upending the presumed radical difference or diametric opposition between them. As memory sites, ruins convey events through multiple temporal layers and evolving uses in memory work, their fullness in the present tied to the "empt[iness] of something palpable in its absence" (Taylor, 2014, p. 242). Not unlike official archives, then, ruinscapes of Holmesburg enact their own politics of domination at the level of both descriptive and redressive documentation and commentary.

Designed by Wilson Brothers & Company, the seventeen-acre compound of Holmesburg Prison was built in 1896 and originally consisted of six sixteen-foot, barrel-vaulted hallways with 450 cells, each measuring only eight by eighteen feet. Four additional cell blocks were later constructed via prisoner labor to ameliorate overcrowding. The entire compound is enclosed on all four sides by concrete barricades as high as thirty-five feet above ground and another twelve feet beneath the surface, giving it the look of an impenetrable fortress straight out of the Middle Ages.[10] Re-creating the radial, hub-and-spoke plan of Eastern State Penitentiary, which opened in 1829, Holmesburg also borrowed its Gothic revival style of imposing crenelated guard towers and a single large central gateway that limited possibilities of escape (figure 3.1). Yet behind this grandiose, medieval façade was a modern prison design philosophy popularized by Quakers in the early nineteenth century. Prior to Eastern State, prisons were more communal, holding captives together rather than apart. Attributing the spread of idleness and corruption among prisoners to this unrestrained contact between them, Quakers commissioned the building of a new prison with individualized cells for solitary confinement and common areas for enforced labor. This new prison, the first of its kind the world had ever seen, was Eastern State Penitentiary, a new system of confinement consciously created for prisoner rehabilitation, stressing the capacity of built environments to improve upon an individual's character.[11] A corrective or progressivist response to what was observed to be an epidemic of moral contamination between prisoners, the combined implementation of labor and solitary confinement—known as the Pennsylvania System and promulgated as "reform" in its insistence on cultivating penitence—is said to have modernized prison management and architecture by focusing (or, to Michel Foucault, *founding*) the individual subject of crime and punishment.[12]

From the outside, Holmesburg Prison appears formidable; it is enormous and sprawling, with thick and heavy stonework. But inside the prison, that impressive image of physical might quickly falls apart; Holmesburg is in tatters, frail and delicate in many places. Photographer Matthew Christopher (2014) argues that this ruined state signals the failure of modernity, which the prison had helped inaugurate. For Christopher, ruins index an "impending social collapse" (p. 7), and he

Figure 3.2. "a means to an end." Courtesy of the photographer. Copyright Matthew Christopher.

Figure 3.3. "a way to make amends." Courtesy of the photographer. Copyright Matthew Christopher.

sees in Holmesburg's history of medical and carceral abuse "an important reminder of who we are, what we are capable of, and how frighteningly close we are to the worst parts of our own past" (p. 123). Though he calls Kligman's medical experiments a "dark passage" in the prison's history, he also situates it as yet "another" instance of normalized violence against its prisoners, which had included beatings and brutal retaliations against strikes. Taken between 2010 and 2014, pictures like "a means to an end" and "a way to make amends"—image titles evoking, respectively, the "worst parts of our own past" and our responsibility for them—attempt to visualize the "particularly barbarous" and "terrible place" that the prison had been. Imaging one dilapidated cellblock, "a means to an end" is a breathtaking display of boundless wreckage and of the massive scale of Holmesburg Prison, its very deep perspective showing profuse corrosion from floor to ceiling far into the visual field (figure 3.2). In contrast, "a way to make amends" focuses on the gated entrance of a lone cell in this carceral universe, the number "1066" flaking off right above it—a nondescript cipher that imparts nothing more than the cell's rather unexceptional, fungible place among many, many other units just like it (figure 3.3). Nicole Fleetwood (2020) argues that carceral visuality is fundamentally a punitive framing, one that symbolically replicates the prison's removal of the subject from the rest of society. Photography in prison spaces is hence overdetermined by the space, its lenses "subordinate to the lenses of the omnipresent watchfulness of carceral optics, even [if] insisting upon other modes of representing the imprisoned" (p. 90). This is present in Christopher's ruinscapes, which insist on connecting the prison space to the free world. Despite the difference in visual scale between his two photographs, both impart the anonymizing effects of prison architecture, the complete absence of particularity wherever the camera is trained: a long hallway that seems to never end, or a single enclosure that copies hundreds of others. Whichever way it frames the space, the camera cannot recuperate singularity from the prison's violence of endless substitution.

In both photographs, the hallway is neither dark nor bright, the sunlight that filters in from overhead skylights producing discontinuous shadows along scattered mounds of rubble and debris and revealing moldering concrete, broken water pipes, and bits and flaps of paint shedding from whatever surface remains underneath. To Christopher, these images of ruin also picture the breakdown of human values along-side the breakdown of institutions, suggesting the indelible workings of dominant ideology within architectural design.[13] Yet in the context of expanding US networks of incarceration at home and abroad, these pictures of Holmesburg cannot help but convey the awesome progress of those same values and institutions by showcasing and archiving one of their most successful prototypes: a relic of the carceral evolution toward what is now called the prison-industrial-complex, whose captive population and capital circulation are unmatched by penal systems anywhere else in the world. Though defunct and outdated, Holmesburg Prison as historical space and as borrowed design concept had been a prosperous forerunner to mass incarceration, its breakdown in real time only rendering it a veritable monument to our prison nation.

As Jani Scandura (2007) writes, modernity and progress are built upon their own refuse and dumps, the increasingly rapid amassing of which indicates the growth and dominance of capital accumulation. This is what makes detritus an unstable category, containing within it a symbolic order, the refusal of that order, and the motivation behind this refusal. Situating archival practices within modernity's dumping grounds, Scandura illustrates how this transformation of refuse into ar-tifacts continually reconstructs what "we allow ourselves to remember—and [what] we force ourselves to forget we ever knew" (p. 8). So while working with remnants purposely thrown away—studying them, col-lecting them, imaging them—may transit between capitalist fetishiza-tion and recuperation, they are also already displaced in both the past and the present, in both time and space. This constitutes the interpretive challenge behind making and reading ruinscapes, the smooth veneer of the photographic image belying the material and symbolic disarray of buildings turned into junk, reaching for unity where none may be found.[14] Does the photographic screen simply record and reflect the un-boundedness of ruin, or does it begin to take the place of architectural skin steadily lost? Christopher hesitates to give meaning and intention to his photographs, reluctant to invest in them a historical and social lesson about ruins. "Rather than the photography of ruins existing for its own sake," he contends, "it must justify itself by what it does or tries to do" (p. 6). In essence, this statement collapses the image with its ob-ject, making the photograph as unanswerable to function and meaning as are the rubble and scraps it captures—an anti-archival impulse.[15] By withholding understanding, by refusing to instruct or offer up a message, ruin photography can also traffic as just more garbage or just more junk propagated in the mass reproduction of loss and memory. This might be read as a denunciation of capitalist principles of purposiveness and productivity if not, as Scandura shows, for the reciprocal relationship between capitalism and its refuse. Thus emptied of meaning, ruin photography reveals the image not as content but as modernity's residual form: waste. The ruinscape does not infer the impossibility of full knowledge from looking or even of the gaze's impossible intention toward unlimited vision. Rather, it implies that even if ruin photography were to imbue its object with meaning and value, it too is seemingly acted upon by the photographic object, its close proximity to or its *emplacement* within ruin and garbage conditioning its own imagistic and interpretive possibilities. There is no need to abstract from the ruinscape, because it is quite literally what it shows.

The aesthetics of ruin, Robert Ginsberg (2004) argues, is fundamentally tied to a loss positioned somewhere between the structure and the photographic image. Ruins compose geographic and temporal sites/ sights of incongruity between presence and absence: "The ruin is the revenge of the formerly unseen upon the whole made invisible. . . . The hidden becomes evident, while what ordinarily is present is made absent" (pp. 34–51). In ruinscapes, this incongruity subverts photography's intentions toward identifying wholes and unities, the camera always failing to capture the aesthetics of ruin by necessarily framing it, segmenting and isolating it into parts—photography *ruins* its object. In ruin photography, the screen of the photograph assumes the broken skin of the architecture, neither possessing a surface upon which meaning can be readily ascertained. To look upon the remnants of vanishing structures is to look upon a certain lack, but where that lack is located—in the ruin, in the photographic apparatus, in the image—is not easily circumscribed.

And this lack is perhaps what is making itself known to Christopher in his own images. It must be noted that his pictures of Holmesburg do not address all forms of ruin and ruination. The problem of meaning which with they grapple does not follow from pictures of epic and celebrated structures like the Colosseum but is instead specifically tied to images of a relatively obscure abandoned prison. Holmesburg just does not possess the same interpellative power that prominent and beloved monuments do. A basis for the free to gaze upon conditions of unfreedom, however, a reason for looking can resist a forgetting of subjugated others. Investing the ruinscapes with purpose shows the reason to look to be as much a reason to *care* to look. Still, as Ginsberg argues, form matters, and the ruinscape's own resistance to wholeness disarticulates the act of seeing from that of knowing, betraying a critique against the elevated status of vision as the privileged route to truth and as the paradigm of Western epistemology.

Those like Christopher who specialize in ruin photography captured more of Holmesburg's spaces beyond its cells, including courtyards, guard towers, corridors, work spaces, and stairwells. True to their preferred genre, these photographers also situate Holmesburg Prison as one subject among others in their portfolios. Christopher's own body of work exhibits abandoned schools, hospitals, churches, hotels, and various industrial buildings; and without context or identifiers, many of these photographs could be mistaken as having the same subject or as having been shot at the same location. Nestled among these other representations of shabby buildings, Holmesburg as photographic subject seems rather unexceptional, constituting just another setting upon which decay and abandonment alight and linger. Ruin photography shows decline to be a quotidian occurrence shattering the historical and formal specificity both of buildings and of their representations. By positioning Holmesburg amid other blighted structures, Christopher's ruinscapes also gesture to the ordinary presence of prison spaces in the American landscape, a golden gulag whose rapid spread Ruth Gilmore (2007) traces to the prison-building boom of the late twentieth century. Occurring soon after the nation's "golden age," the beginnings of mass incarceration were coeval with a burgeoning throw-away economy including both unwanted commodities and expendable people. Taken up as geographic solutions to the social problems and political crises brought on by the wars and national and global liberation movements of the 1960s, prisons also aided the economic recovery of rural sectors still ailing from the Great Depression by funneling into them the growing captive population of urban cities. Prisons transported jobs and capital to places the New Deal could not recuperate, these politically neglected regions becoming dumping grounds for discarding communities made disposable in metropolitan centers—differences that, as Gilmore writes, had already been "emblazoned on surfaces of skin, documents, and maps" (p. 15).

Presenting the ruins of Holmesburg Prison alongside that of schools, factories, hotels, and other mundane buildings, ruinscapes like Christopher's visualize captive sites as the familiar structures of daily life that they have become. Across the photographs, ruin thematically connects and binds together disparate locations and edifices, blurring and distending the architectural and operational differences between types of built environments. Between the picture and its object, ruin cuts both ways; and if obsolete, dilapidated buildings are made equal in their status as refuse, then ruin photography becomes an apt genre for banalizing rather than spectacularizing the prison space. Michel Foucault had long discerned the network of power running through various institutions like the clinic, the prison, the church, and the schoolhouse. But as ruinscapes perhaps demonstrate most pointedly, it is through the mounting refuse of these institutions that the presence of power and its interconnected, ordinary workings become markedly perceptible: the logic of captivity pervading the everyday, in continuous recycle. Holmesburg Prison was decommissioned in 1995 to move its operations to the newly constructed maximum-security stronghold Curran-Fromhold, which is currently Philadelphia's largest correctional facility. Like the disposable commodities of late capitalism, Holmesburg Prison modeled and facilitated its own replacement by a bigger, better, and sleeker stockade of human confinement, its ruin thus signaling not the failures of incarceration but the successes of its mass reproduction, the timeless rubble of eternal catastrophe.[16]

The Skin of the Architecture

What to make of ruin and ruinscapes as archives of Holmesburg's history of medical experimentation on prisoners? How to read this history in photographs whose flat, even surfaces quite literally gloss over the roughened, dislocated contours of what once had been a carceral and experimental space? Even when accompanied by textual accounts of what used to happen at Holmesburg, the photographs do not record the time of Kligman's experiments. Rather, the rust, debris, and layers and layers of exfoliating façades picture a stratified temporality of past and present, with ruin photography becoming a visual "eulogy" that, Christopher argues, transforms ruins into the corpses or "bodies of hopes and ambitions, and in their link to our shared heritage and common past, [as] a part of our 'extended family'" (p. 5). To Christopher, ruin photography is not about garbage but about current and imminent societal collapse—"an age of consequences"—as well as a personal "death that awaits us all . . . the frailty of the human condition" (pp. 6–7). Like a eulogy, ruin photography is testimony of a time gone by that serves as lesson or inspiration for those still living, a usable past for an imagined future foreshadowed by a dead or dying body that is both social and individual, public and private.[17]

Ruin as allegory of death is refuse made meaningful for cultural memory, garbage repurposed for new symbolic ends, bringing ruin photography closest to that "agent of Death" Roland Barthes (1981) had first associated with picture taking.[18] Furthermore, ruin photography conforms to Western architectural frameworks that have frequently likened buildings to human bodies. Juhani Pallasmaa (2005) notes, for example, that the "most archaic origin of architectural space is in the cavity of the mouth" (p. 57).[19] However, for Pallasmaa, gustatory, sonic, and visual sensations ultimately extend from one particular sense organ he considers the most dominant: the skin, or "the eyes of the skin," insofar as it constitutes the membranous edifice within which other sense organs operate. Mediating the body's encounter with spaces, the senses—and touch most especially—help project the self into a building, which "functions as another person, with whom one unconsciously converses" (p. 64). The skin trope in architectural theory is also evident in the late nineteenth-century work of Adolf Loos, who is widely credited with inaugurating modern architecture's preoccupation with "clean," unadorned surfaces, later epitomized in Le Corbusier's trademark white-walled buildings.[20] Locating skin as the origin of built environments, the Loosian theory of architecture centers the direct communion between skin and what it protects or covers like fabric, evincing a "desire to *house* the body [that] grows most vitally out of the desire to *be* the body" (Cheng, 2011, p. 54, emphasis in original). Evoking the loss or decay of skin, ruins are thus unsurprisingly analogized to dead bodies

Figure 3.4. "Cell 560," 2011. Courtesy of the artists, Patricia Gómez and María Jesús González. www.patriciagomez-mariajesusgonzalez.coml; http://www.philagrafika.org/

Figure 3.5. "Cell 560," 2011. Courtesy of the artists, Patricia Gómez and María Jesús González. www.patriciagomez-mariajesusgonzalez.coml; http://www.philagrafika.org/

whose decomposition is made most visible through the gradual loss of barriers protecting and cohering inside from outside.

This conceptual transition from skin, fabric, body, and architecture, as well as a critical commentary on incarceration, is present in the multimedia work *Doing Time | Depth of Surface*, by Spanish artists Patricia Gómez and María Jesús González. Commissioned in 2011 by the Philadelphia-based printmaking collective Philagrafika and exhibited in the same year at the Galleries at Moore College of Art and Design, the artwork sought to conserve and produce large-scale prints of Holmesburg Prison's fading graffiti and posters put up by those it formerly caged. To do this, Gómez and González employed a restoration technique called *strappo*. Their modified *strappo* procedure involved applying water-based glues to prison walls and then painstakingly stripping and transferring the surface paint onto black cloth canvases, resulting in stunning monoprints that, to paraphrase curator José Roca, transformed the skin of the architecture into crumpled shrouds (figures 3.4 and 3.5).[21] But although death is implied in the work's aims at preservation, it is life or traces of prisoner activity—the "depth" to the surface of walls—that animate Gómez and González, who had previously deployed similar strappo techniques in prisons at Valencia and Palma de Mallorca, Spain.[22]

Print, specifically "large-format print realized without a press, ink or paper," is especially conducive to salvaging captive stories (Gómez & González, 2011a, p. 15). The artists explain:

> For us, the connection between human skin and architecture is as critical as it is obvious . . . the walls inside a prison where an inmate expresses himself are like a *second skin* that envelops and protects him, separating him from the exterior but also imprisoning him. When nothing remains of a place and its walls are the sole element left to tell a story, our job is to reclaim and reveal those histories" (ibid., emphasis mine).

However, the process of reclamation is as precarious as the histories it attempts to save. Unlike ruin photography in which one can keep taking pictures as desired, Gómez and González's printmaking technique does not allow for second chances. *Strappo* permits no attempts beyond the first, with failure resulting in the permanent destruction of artifacts and the very loss, both physical and metaphysical, that it strived to defer. But

Figure 3.6. Photograph of completed canvas, laid flat in B Block corridor of Holmesburg Prison. 2011. Courtesy of the artists, Patricia Gómez and María Jesús González. www.patriciagomez-mariajesusgonzalez.coml; http://www.philagrafika.org/

when successful, *strappo* produces a physical chronicle of captive spaces, "mural membranes" as alternative archives of Holmesburg Prison.[23] For the artists, the walls of Holmesburg Prison composed a sort of lithographic object engraved with "the passage of time," its "historical, social and sentimental information" all indexical marks "on the verge of being lost" (ibid.) but *which strappo* can physically rescue and safeguard from crumbling wall supports.

This more direct, tangible engagement with Holmesburg Prison may attribute to the resulting monoprints a greater sense of historical authenticity by retaining original components of the ruined facility.[24] Still, *strappo* presented its own methodological challenges to a perfect relocation of paint and graffiti. Mainly, the flaking and crusted surfaces of prison walls had to be flattened out—de-texturized, as it were—with surfactants and glues before they could be removed. This definitive element of *strappo* chemically changes the paint while also undoing some of the "magical effects of time" the artists hoped to conserve. Surfaces not entirely flat or unsuccessfully peeled off form blank spaces on the cloth canvases in much the same way that doorways and windows would. Both are preserved as void in *strappo*'s rendering of 3D space onto 2D textile monoprints (figure 3.6). In this way, *strappo* is *fabric*ation in every sense of the word, from building to fiber transposing physical depth into one metaphorical: the "people on the walls of places" (ibid.). What Fleetwood calls "carceral aesthetics," monoprints do double the representational work, generating new creative objects from existing prison art while also performing the recuperative functions of archiving. Indeed, relating to those who had been incarcerated at Holmesburg through the wall art they produced, surfacing hidden cultural expressions almost forever lost, monoprints not only explicitly tie aesthetic judgment with historical recording.[25] They also make visible the carceral geographies underlying the public life of aesthetics and history, forcefully connecting these divided spaces into a single visual field.

Strappo enacts creation and loss as interdependent operations of archival and documentary practice, producing a tension between, to borrow from historian Michel-Rolph Trouillot (1995), "what happened and that which is said to have happened" (p. 3). Combining artistic production with historical preservation, Gómez and González's monoprints stage the meaning-making processes inherent to archival work, wherein preservation produces both something lost and something new. From the walls of Holmesburg Prison to the museum gallery, or more generally, from the site where an object is first discovered to the institutionalized repository where it lands, the archival impulse falls short of the ideal recorder. Turning things into records or *source* materials marshals new uses and interpretations, a network of matter, performances, and significations not the same as but also not quite fully removed from the objects' original contexts (what counts as "original" is also reconstruction). Visualizing this moment between fact assembly and fact creation, when the archival object is both the aim *and* (faulty) medium of recovery, Gómez and González's monoprints are simultaneously originals *and* copies, wherein object and representation are self-same.

To remark on prison time, the artists transition from visual-haptic mediums to one wholly sonic. A four-hour audio recording states again and again a single, recurrent entry in the prison's logbooks: "All appears normal." Tonelessly recited by a former guard at Holmesburg and replayed in forty-five-minute loops, this phrase uttered over and over conjures the rigid schedules and monotonous routines of prison existence as well as its moral acceptability—it is, as the recording repeats, normal. While deterioration captured by monoprints and ruinscapes suggests a linear passage of time, the audio recording evokes the repetitive, circular temporality of captivity, its continual, uniform sound arrhythmic, flat, an unbroken phonic skin eliciting a sensuous history markedly different from that of the visual works. The change in medium invites distinct somatic engagements, separating out the chronological time of decay from the timelessness of prison routine. Significantly, ruinscapes, including those also taken by Gómez and González, image this repetition. Compositionally, their photographs of Holmesburg's individual cells do exhibit a relentless constancy, each a somber portrait of an individual cell unremarkable from all the rest inside the prison. The same form persists in Christopher's collection as well. Nearly identical pictures of cell rooms, cell walls, and cell corners present across their works, often repeating the same perspective. Each cell is, of course, differentiated by its singular manifestation of ruination. One shows a dusty book propped on an equally worn-out cot (figure 3.7), another a barren room minus a large tree trunk bisecting the back wall (figure 3.8), another a rotting metal bedframe in the center (figure 3.9), still another a dim, colorless hovel whose walls are lined with more fibrous vines and roots (figure 3.9). Yet across these photographs, scenes of ruin structurally never change, the narrow selection of framings enacting the same setting over and over again. Specificity escapes the eye, the unceasing repetition mirroring the bleak and severely regimented organization of prisons spaces.

At Holmesburg, a single cell is uniformly replicated multiple times inside a cellblock, the block then repeated nine more times around a panoptic center, and the entire facility constituting one copy among hundreds of Eastern State Penitentiary. In Christopher's and Gómez and González's ruinscapes, form and content are as oppressively unchanging as the prison architecture they image, suggesting that the problem of meaning behind their pictures may not ultimately derive from the disharmonious edifice of Holmesburg's ruin but from a structural fungibility that confounds the meaning and visuality of space, place, and location.[26] Re-creating the excesses of prison architecture, the ruinscapes cannot themselves be but bewildering and superfluous, a visual impasse

Figure 3.7. Photograph of cell 843 interior in I Block of Holmesburg Prison, 2011. Courtesy of the artists, Patricia Gómez and María Jesús González. www.patriciagomez-mariajesusgonzalez.coml; http://www.philagrafika.org/

Figure 3.8. Photograph of cell 844 interior in I Block of Holmesburg Prison, 2011. Courtesy of the artists, Patricia Gómez and María Jesús González. www.patriciagomez-mariajesusgonzalez.coml; http://www.philagrafika.org/

Figure 3.9. "the dream of release." Courtesy of the photographer. Copyright Matthew Christopher.

Figure 3.10. "'Death as the destruction of all things no longer had meaning when life was revealed to be a fatuous sequence of empty words, the hollow jingle of a jester's cap and bells.'~Foucault." Courtesy of the photographer. Copyright Matthew Christopher.

between sensing and sense-making. If we take seriously prison ruin as photographic form, we might encounter its defiance of symbolic intentions or the prison's obliteration of pictorial gestures toward the particular and wonder if the free can ever understand unfreedom when they see it. This troubled visuality reappears in Gómez and González's sound installation, where the prison becomes a flight from the visual entirely. Architectural redundancy presents along another sensorial experience, that of hearing instead of seeing. Yet the spatial logic of captivity persists across media attending to each, connecting the time of the prison to the time of archival art.

The recurring display of prison cells in these ruinscapes is reminiscent of another space organized in like manner: a laboratory's vivarium. A vivarium is a place for keeping, breeding, and managing a relatively homogenous population of live specimens for the purposes of scientific observation and research. Containing cages of the same size and shape, all arranged into a grid-like structure, the vivarium's uniformity in design does not simply constitute an efficient system of warehousing; it also standardizes the specimens by providing each one with the exact same survival conditions. Reliably supplying Kligman's research with a homogenized pool of test subjects, Holmesburg Prison was not unlike a vivarium. The ruinscapes' photo array of prison cells harkens to this standardizing effect of prison architecture, picturing the spatial dimensions of a knowledge apparatus precluding individuality. Making captives as interchangeable as the pens separating and interning them, the prison space, to use Fanon's words, effectively generated two separate human species, here one made for the study of the other. As Donna Haraway (1989) argues, the singular function of specimens is to model life processes not their own, corporealizing the "virus-like repetition" of laboratory spaces. Specimens enact the laboratory's unparalleled efficiency in hijacking and mechanizing vitality and vitality's relational forces, making life imitate life. For instance, researchers study cancer in rats not because they are concerned with cancer in rats but because rats can model or sufficiently resemble cancer in humans. Representing life not their own, specimens are bodies doubly copied, equivalent to each other and to a subject they are not. Topographic redundancy is the built environment of this viral replication of specimens that is, significantly, a reproduction without kinship. Through an entrenched social machinery ensnaring particular bodies for accumulation, the prison reproduces this genealogical violence of the vivarium, which forecloses social relations between specimens and between specimens and those they model. Enrolled in studies generally addressed to the free, the "inmate" test subject, like the specimen, embodies this dual effacement of personal and social identity.[27] And today, cultural references to prison research programs frequently evoke the vivarium when, for example, they liken captive test subjects to "human guinea pigs." Thirty years after Kligman's program closed at Holmesburg, a lawyer for several hundred former test subjects noted, "Today, more consideration is being given for lab mice and rats than was given [to] human beings."[28]

The term *specimen* contains within it visual resonance, combining the seventeenth-century Latin terms *specere* (to look at) with *men* (a result or means). Specimens are both instruments and objects of the scientific gaze, comprising the products and the modes by which those products come into view. Hence, specimens are not per se the subjects of the gaze, because there would be nothing of scientific significance to look at if they were looked at as subjects in themselves. And yet Christopher's and Gómez and González's ruinscapes do just that: subjectifying the specimen by calling attention to the spaces that define it. If, as architectural theory illustrates, aesthetics and built environments are created with certain bodies in mind, then what takes shape in their photographs of prison cells is a subject that throws subjectivity itself into crisis.[29] Rejecting the metonymic function of the specimen, ruinscapes create a visual terrain where the specimen undoes the same forms and objects of looking that it enables. Like large monoprints, the ruinscape is a visible individuation of prison cells, each transformed into singularly evocative art.

The large-scale monoprints, however, are the centerpieces of Gómez and González's exhibition, receiving the most critical attention. Critic Jennie Hirsh (2011) highlights the multiple histories present in the layers of salvaged paint and the dual levels of recovery at work in this layering: first, the mark-making practices of prisoners, their graffiti, as a form of resistance against their abject status, or as a defiant recuperation and re-presentation of subjectivity; and second, the monoprints as archival process seeking to bring back and have us bear witness to remnants of lives brought out of the time and space of civil society. To Patricia Robertson (2011), the "individual and unique" prints collaborate with the past to preserve "vital experiences," forming "pictorial analogs for the singular lives . . . that unfolded in [Holmesburg's] uniform spaces" (p. 18). Indeed, the prints are seen not only as evidence of what prisoners did to the walls but also as counterparts to the prisoners themselves, as visibly standing in for their absence, including those who had been subject to Kligman's medical experiments. Kostis Kourelis (2011) writes, "We seek earlier occupants, such as the infamous inmates who received dermatological tests. . . . Although we have no physical evidence of this event, [Gómez and González's] project allows us to imagine the invisible shadows of such powerful interactions between architecture and its users" (p. 2). Writing for *Art Papers*, Edward Epstein (2012) views the monoprints as a "commemoration of this act of medical hubris," a "limp enclosure" akin to "animal hide" that provides a "fitting tribute to what Kligman himself describes as the 'acres of skin' to which he applied his untested chemistry" (p. 52). And as Gómez and González also say of the prison, "What has happened to Holmesburg's walls echoes what happened to the skin of the inmates who participated in medical experiments that had grave health consequences" (p. 15).

Such observations use the body of the prisoner to anthropomorphize the prison, humanizing the structure through analogies made between corporeal and architectural injury. But these descriptions also point to the status prisoners occupied as objects, represented not only through their inscriptions—graffiti, posters, markings, decorations—but also through the very stone and concrete that barricaded them from the rest of society. By inviting viewers to see in Holmesburg's disintegrating walls the violated bodies of Kligman's captive test subjects, Gómez and González's monoprints enact Holmesburg Prison as a site/sight of both subjugation and subject-formation, as the means for reading the concurrent making and unmaking of captive being. The artists describe their monoprints as "giving voice" to those captives, these reconstructed prison walls becoming by proxy the active (empowered?) agents of telling captive experiences to spectators in the free world.[30] Mapping the skin of prisoners onto the skin of architecture, Gómez and González's monoprints betray the authority, the representational power, of those walls to signify those they locked up, the latter becoming known and remembered through prison walls that talk, prison walls given voice by those they kept captive.[31]

This transfer of agency and interiority from captive bodies to the structures that caged them recalls the powers of "living death" Dennis Childs (2015) ascribes to the prison's capacity and intention to "immobilize, torture, and kill" (p. 35), powers acquiring for the prison a "life of its own" by siphoning and diverting to itself the social and biological life of the captive (ibid). Archiving and memorializing captive subjectivity through the walls of Holmesburg Prison, Gómez and González's monoprints re-spatialize this displaced subjectivity, materializing the dual status of captive spaces as simultaneously sites of "black vernacular cultural production" (p. 22) and "the sepulcher-like temporal boxes of the master archive" (p. 6). Making seen the unseen residues of captive life, transplanting artifacts normally concealed behind prison walls into the gallery's specular spaces, the monoprints may confer upon Holmesburg's graffiti the status of high art, unsettling dominant notions of art aligning aligning it with the free. They may reproduce hierarchies of power by fetishizing the object of the Other in a museum setting Michelle Wallace (2002) has called "the prison house of culture."[32] Producing a carceral aesthetics, monoprints may serve abolitionist interpretations reconfiguring another kind of public inclusive of prisoners as makers and viewers of culture. They may signal the making of alternative archives as well as the limits of reaching, representing, and preserving repressed histories. Imparting a panoply of textures from Holmesburg's flaking, fragile surfaces, these prints may contribute to what Laura U. Marks (2000) terms a "haptic visuality" or touch epistemology, which emphasizes the always embodied, multisensorial nature of making and viewing images, and the always tactile, sensual quality of memory. They may be all of these and more, but what they pointedly show are the ways in which history is found or located not only in the past but also in present projects of meaning-making.

The tenuous distance between Holmesburg Prison as captive *site* and Holmesburg Prison as captive *body* not only sheds light on the ambiguous relationship between inside and outside introduced by skin, which is permeable and thoroughly bound up with what it keeps within. It also obtains in relations of domination at the fraught intersections of redress and appropriation. Like the muddied borders between prison architecture and imprisoned body, the representational slippage between ethical accountability and the exercise of power reproduces the captive body's inability to draw lines around itself. Recognition and disavowal—or, from Christopher's ruinscapes, a way to make amends or a means to an end—both make visible the unconditional availability of the captive condition to meaning- and memory-making. Skinning the body of prison architecture conceived as the corporeal body of the captive, monoprints aestheticize the latter's absence of physical and symbolic edges, able to be both surface and depth of representation.

Skin as Ghostly Matter

"Nobody has come in here to investigate. We're the first ones! Slam one of these doors! Now's your time! Now's your time! If this is a portal to hell, slam a door! I'm trying to honor you the best I can even though when you were alive, you did terrible things." It's near midnight, and Chad Lindberg, a paranormal investigator, crouches on the floor of Holmesburg's central rotunda while pleading with spirits to commune with him. Alone in the deserted prison, pitch black at night (figure 3.11), Lindberg can see only with a single handheld flashlight that he aims intermittently at the entrance to one of Holmesburg's ten corridors: H Block, once the center of Kligman's medical experiments. Sensing no signs of ghostly presence, Lindberg continues: "But it can be made better! You can cross over!" He pauses, straining to see or hear any changes in the dark silence of "the terror dome," a nickname guards and prisoners had given the vestibule. Then, somewhere behind Lindberg, a clang rings out, like the sound of a metal gate roughly shut. He screams and scrambles to his feet, turning and pointing his flash flight at the direction of noise. He sees nothing but keeps screaming anyway. Later, once calmer, he approaches one of two static cameras taping the scene in night-vision and says to the viewer, "I've officially . . . ," he shudders and looks behind him, "I've officially lost my fucking cool." In voice-over, Lindberg explains that the sound is without a doubt from a cell door slamming shut off-screen.

Imbued with narratives of death and decay, the ruins of Holmesburg Prison are, perhaps unsurprisingly, generative ground for paranormal investigators like Lindberg. Lindberg cohosted the 2014 Destination America miniseries *Ghost Stalkers* with John Tenney, both visiting Holmesburg in the show's fourth episode. Airing for only one season,

Figure 3.11. Night-vision view of Lindberg from a static camera located in the Holmesburg Prison rotunda. From season 1, episode 4 of *Ghost Stalkers*, 2014.

the series follows Lindberg and Tenney as they "spend forty-eight hours isolated in some of the rawest, grittiest haunted locations in the world," locations often either abandoned or converted into historic sites. Lindberg and Tenney's fascination with haunted places began with their own near-death experiences. Their narrow escapes from death in youth gave rise to a long preoccupation with bridges to "the other side." The "mission" for these cohosts of *Ghost Stalkers* was to discover such "portals where the dead can cross over into our world." Holmesburg's hub-and-spoke design holds particular interest for Lindberg and Tenney, its "unique" construction possibly making for a man-made portal generator: "Its multiple hallways converging beneath a parabolic dome could've created a situation where a century of negative human energy was [shunted to] and trapped beneath the central rotunda . . . tear[ing] open a gateway to another realm." To monitor the electromagnetic energy ostensibly given off by portal activity, Lindberg and Tenney install a "wormhole detector" beneath the dome.

Arguably a cast member unto itself, the wormhole detector figures prominently in the *Ghost Stalkers* series, its entrance into every episode marked by careful setup and explanation. Representing "scientific methods and cutting-edge technology," the detector converts each supernatural encounter into empirical investigation. It registers fluctuations in gamma and electromagnetic radiation, changes ostensibly caused by the presence of portals or of spectral beings passing through them. The exact location of a portal is ascertained through four sensors strategically positioned inside a haunted space, their detections of varying radiation levels graphically monitored on a laptop computer (figure 3.12). On screen, paranormal activity appears as large spikes of radiation resembling seismograph readouts (which measure the strength and duration of earthquakes). The more sizable and numerous the spikes, the closer a portal or a ghost is to the sensor. The detector is then able to quantify and locate in the real world a phenomenon presumed non-existent, transforming the stuff of superstition and mysticism into verifiable objects of specialized knowledge and professional practice. And more than any other instrument deployed by cast members—handheld cameras, static cameras, sound recorders—the wormhole detector provides the most definitive proof of supernatural activity and is frequently made to corroborate evidence captured by these devices. When Lindberg "loses his cool" in the rotunda, for

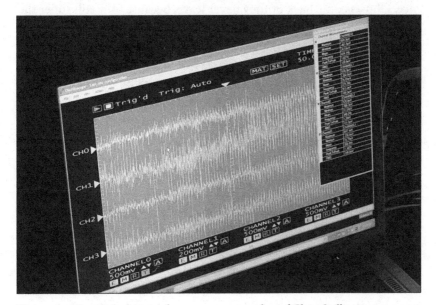

Figure 3.12. Wormhole detector from season 1, episode 4 of *Ghost Stalkers*, 2014.

example, the wormhole detector displays many large spikes denoting changes in radiation levels nearby. This announces the presence of something or someone beyond direct experience. Audio data only hints at this paranormal activity—the slamming of a cell door—while visual devices did not capture it at all—both the slamming and its cause. But the wormhole detector could not only sense the activity but also represent it in graphic form, which becomes the basis for estimating the location of ghostly arrival. Clearly outperforming other technologies of observation and recording, the wormhole detector is key to the science of paranormal investigations that in *Ghost Stalkers* are no longer just the peculiar interests of occult enthusiasts but the demonstrable knowledge of trained experts.

At first blush, Lindberg and Tenney appear less interested in Holmesburg Prison as they are in the possibility of portals, devoting much of the episode to discussing and searching for this space beyond space. This is ultimately what they hope to discover, not where the spirits of the dead reside or where the living still endure but the doorways that connect those worlds together. Able to pass through these doorways, ghosts are signposts of portal activity, their presence indicating a gateway to the underworld wrenched open, usually, by traumatic experience. Ghosts come back or they never leave, because they still suffer the painful events that led to their demise. Interviewing former guards at Holmesburg, Lindberg and Tenney relate the prison's "lurid history of abuse, murder and mayhem": in 1970, a riot that left nearly one hundred prisoners injured; in 1973, a surprise knife-attack killed both the warden and deputy warden in their offices; and in 1938, the infamous "Bake Oven Deaths," which saw four prisoners trapped inside a cell for fifty-eight hours while it reached temperatures of 190 degrees.[33] Described as "hell on earth" and "the worst that humanity could offer," Holmesburg seemed the ideal place for Lindberg and Tenney's paranormal research, rife with the vengeful spooks and hair-raising haunts that would entice self-described "ghost stalkers." Tellingly, access to the site was fairly easy to obtain, as Tenney writes, "not as hard to do as [with] other locations *since it's a prison*."[34] He echoes the ease with which postwar medical researchers had acquired their test subjects, intimating that, even in its afterlife, Holmesburg remains a locus for scientific knowledge production, dubious or accepted.

Documenting journeys to haunted sites, *Ghost Stalkers* participates in a wider economy of commercialized voyages to the macabre called "dark tourism," which Tiya Miles (2015) describes as the "the exploration of death, disaster, and suffering through travel" (p. 8). Miles studies dark tourism in the US South, whose booming industry of ghost tours in plantation mansions combine antebellum history with narratives about the lives and deaths of slaves who are now claimed to haunt them.[35] Fair game to dark tourism's commodification of history and bad feeling—the pleasure in experiencing negative emotions like fear and anxiety—plantation ghost tours speak to the continued profitability of the enslaved and to the myriad commercial and affective uses of black subjects even after death.[36] Like these ghost tours, *Ghost Stalkers* also purveys sensationalized captive experiences, selling and marketing sites of trauma and dispossession as science entertainment or, in Miles's words, the "lighter side" of death.[37] Yet Miles hesitates to dismiss ghost stories as mere frivolous fun, arguing that however much they tame or appropriate the troubled histories they present, these narratives not only afford a way of accessing the past; they also provoke a consideration of how the past is accessed at all, how the stories we tell about the past also retell our uneasy relationship with the present. Ghost stories make up "fringe" histories, "popular forms of historical narrative" (p. 15) that through their retellings demonstrate the *unsuccessful repression* of painful cultural memories. Much like more accepted forms of history-making, ghost stories cannot help but reenact the power and politics they conceal, betraying how the act of concealing, itself, inevitably constitutes a revealing of something else.

In *Ghost Stalkers*, this something else could be what Sharon Patricia Holland (2000) calls "the burden of the national id" in cultural imaginations surrounding both death and black subjects. Narrating abuse, murder, and mayhem or the hell on earth that could only come from the worst that humanity could offer, *Ghost Stalkers'* episode on Holmesburg Prison projects onto its captive dead the id's nature of unbridled aggression, destructive drives, and suspended morality, implying the more ordered and controlled activities of not only the living but also the free. As agents of the living and the free, Lindberg and Tenney enter Holmesburg as if they are paranormal prison guards regulating the wanderings of captive ghosts, their wormhole detector a surveillance apparatus sensing beings-out-of-place who can then be made to "cross over" or return to their proper ground. This inverts the relationship between research and incarceration established during postwar medical science. Whereas medical researchers had capitalized on the prison's controlled environment in the past, today's paranormal investigators like Lindberg and Tenney use their science to reestablish the control lost, ironically, through a prison architecture conducive to wormhole generation and the errant dead. By showing up where they should not be, ghosts are by definition criminal, and science once more enters the prison to help manage and mollify disturbance.

To read *Ghost Stalkers* in this way or, more generally, to take ghost stories as seriously as they take themselves occasions a different interpretive practice that Miles calls "ghostwriting" or what Holland means by "raising the dead." For Miles, ghostwriting calls forth the power of *real* ghosts, which are not simply the friendly or hostile spirits encountered in tours and popular media. Rather, these real ghosts are the unanswered political questions and stakes, or the "deadly serious messengers," of the past implicated in every act of (hi)story-telling (p. 132).[38] And for Holland, raising the dead is critique that transgresses the boundaries between belief and knowledge, life and death.[39] This is precisely the transgression that ghosts enact and that makes critique another doorway through which ghosts can appear. Thus, pace Miles and Holland, *Ghost Stalkers* cannot be simply written off as shallow entertainment or bad ideology, even if they may well be those, too. If the ghosts of Holmesburg Prison are real—if they are deadly serious messengers from the past—then the means through which cast members attempt to register their presence also require serious examination. Claims of pseudoscience would lead critique off-track, not only because they institute a binary between true and false knowledge that both Holland and Miles caution against, relegating the ghost to mere illusion. They also disregard the elements of myth, belief, and superstition that at least one celebrated philosopher of science had observed in the making of accepted or paradigmatic scientific thinking (Kuhn, 1962/1970). Who can argue that the black box of science and technology does not in its own language, in its own "enchanting discourse" (Wynter, 1987), mystify the workings of nature? Taking Holmesburg's ghosts seriously, then, means regarding *Ghost Stalkers* as a popular historical narrative and televisual documentary of the science of paranormal activity. And it also means tracing their unsuccessful repression of painful cultural histories or their drive to banish the ghosts haunting our accepted archives.

A readiness to take the ghost seriously, however, does not mean that the ghost is readily accessible. In the *Ghost Stalkers* episode on Holmesburg, ghosts actively thwart visualization by interfering with media technology. The ghosts are said to be constantly rendering inoperable the static cameras placed throughout the prison, bewildering Tenney and Lindberg during each of their solo overnight stays. While walking through H Block, for example, Lindberg becomes alarmed when the static camera placed inside the hallway suddenly shuts down. He exclaims, "*Oh* shit! *Oh* shit! My camera battery *just fucking died*. Just fucking died!" Training his handheld camera to the one now obsolete sitting atop its tripod, Lindberg exclaims, "Look at that! I had a full fucking battery when I came in," remarking snidely to the ghost responsible, "Okay, well now you should be charged up for sure!" In voice-over, Lindberg explains that ghosts absorb energy from electrical devices in order to "manifest," or to "move objects, make noises, or simply make their presence known." Not long after, Lindberg must also change out the batteries of his handheld camera, the fully-charged ones it began with already drained of power in just fifteen minutes of use. Such rapidly depleted video camera batteries indicate to Lindberg that "whatever" was in the prison with him seemed to have "an insatiable appetite for energy." This energy coming from visual instruments animates the ghosts of Holmesburg, who manifest by appropriating or meddling with modes of witnessing their manifestation. Mechanical forms of documentation (cameras) succumb to ghostly appetites for powers of observation (energy), their capacity to record and provide visual evidence foiled by their own intended subjects of the look. The spectral and the ocular here inhabit conflictual positions, any knowledge gleaned therein retroactively derived from sabotage; we know the ghost is there because it is eluding us. It is most palpably present when there is *lack* of visual proof. Avery Gordon (2008) says this is the ghost's "visible invisibility: *I see you are not there*" (p. 16, emphasis in original). But visible invisibility, at least at Holmesburg, is not the essence or inherent nature of ghosts, emerging only through their parasitic interactions with visual technology. As the work of hungry ghosts, dead camera batteries point to appearance without image—indeed, to appearance that consumes the image, killing it, producing what Michael Taussig (1993) terms "*pure appearance*, appearance as the impossible" (p. 135, emphasis mine).[40]

Aside from abrupt electrical failures and eerie sounds with mysterious origins, Holmesburg's ghosts do exhibit a strong preference for one mode of communication in particular: touch. If ghosts are not seen, they are certainly *felt*. Throughout the episode, both Lindberg and Tenney can be seen jumping or recoiling in fright, startled by invisible caresses, taps, and pinches. "Who's touching me?" Lindberg blurts out in the dark empty room of the prison cafeteria during his exploration, alarmed by a slight, unseen brush against his forearm. That question must have displeased the offending ghost, who strikes (kicks?) Lindberg's shin. "Ow! Ow!" Lindberg hisses in discomfort, clutching his leg and rubbing the sore spot. Perturbed but undaunted, he repeats a common refrain in the episode: "In what part of the prison is the portal?" Hearing no reply, he draws his arm closer to his body, sighing in discomfort, "*Ahhhh* . . . it's like they're tugging on my . . . on my arm hairs. Just *tugging*." In Tenney's own turn inside the cafeteria, "something or someone" also makes contact. Hearing and "feeling" people all around him, Tenney becomes agitated when he feels a "really intense" touch at the back of his head. "Oh!" he shouts gripping the area, "I got touched! I got touched! Something just came up behind me!"

Sensing ghosts through the skin, both Lindberg and Tenney are themselves akin to apparatuses of detection, instrumentalized by their own subjects of study. The surfaces of the body, not the eye nor the camera, become the privileged medium of ghostly expression, transferring authority over knowledge production from paranormal investigators who see nothing to the ghosts who make physical contact at will, touching while being themselves untouchable. Having lost control over their research subjects, the scientists are denied full knowledge. Neither Lindberg nor Tenney ever lay eyes on the ghosts they encounter, but they know the ghosts are there because they can *feel* them— knowledge is quite literally superficial. Like their cameras, however, Lindberg and Tenney also risk being drained of vitality. Before his solo tour of the prison, Tenney says of it, "This place is sucking every ounce of energy out of everything, *including me*." And when he later enters H Block, he does come close to collapsing, very nearly fainting from "something—anxiety, fear, a tightness and pressure growing in [his] chest," a something he frantically attempts to escape as he flees from inside the hallway. Calling for help, he stumbles and falls to the ground just outside of H Block under the terror dome, where Lindberg later finds him crumpled, out of breath, and begging to be let out because he was, in his words, "cashing out." As with cameras, the bodies of cast members are sources of energy for hungry ghosts. Though proffering a more visceral encounter with them—visceral in both the corporeal and affective sense (anxiety, fear, a tightness and pressure in the chest)—this loss of power which allows the body to feel the ghost or which brings the ghost in closer proximity nevertheless constrains that body's capacity to apprehend it further. Or better put, it is precisely these constraints to apprehension, turbulent sensations like anxiety, fear, difficulty in breathing—all signs of a panic attack—that make of haunting an object not of cognition but of felt experience.[41] This is what makes Holmesburg's ghosts frightening, not because they cannot be seen or because they are effectual despite this. They are frightening because they contravene any possibility of mastery, demanding that knowing come from a place of vulnerability, uncertainty, and loss of control. Ghosts exact the same price from cameras as they do from cast members: the authority to represent or symbolize their deadly serious messages. Communicating through the skin, these famished ghosts shift from pure appearance to pure surface, surface as the impossible, where embodied knowledge means disturbing the knower's relationship to their body by relinquishing the body to the ghost.[42] Wormhole detector notwithstanding, the ghost will not be scientized.

Ghostly resistance to mechanical visualization and human perception produces disorientation as the felt experience of hauntings, and this is evident in the episode's dialogue on the precise location of wormholes.[43] Cast members oscillate from uncovering them *within* the prison to situating *the entire prison itself* as a passageway between life and death. "I think *this*," Lindberg says of Holmesburg, "is a portal to hell. I think this *is* hell. I think we are in it *right now*." Tenney concurs: "I can tell you from someone who might have been there, it's the closest thing on earth to it." Among all the haunted places he has surveyed, Holmesburg Prison is special to Tenney because it has brought him the closest to his near-death experience. Recalling this experience, he says, "You know, that place I went to when I died was what I think hell was probably like, and prison—being locked into a place for the rest of your life—that's a hell on earth." Michael Hardt (1997) calls imprisonment an "exile from the time of the living," a "wasted, impoverished existence" that reduces captives to "mere shadows" pushed out from their very bodies (p. 66). Though biologically alive, the captive is already a specter, already at the interface between life and death. Following their close encounters with ghosts, Tenney and Lindberg unwittingly imply this true form of the prison, in their words, a "purgatory" that will forever remain "here on earth" for as long as Holmesburg(s) exist. Ironically, they lost the doorway they were looking for as soon as they had set out to find it—with their cameras and detectors—rediscovering it only after they had been beleaguered by ghosts during their solo tours or only after having floundered through haunting's disorienting effects. At the episode's closing, Lindberg and Tenney remind the viewer what they had already noted at the beginning: that Holmesburg is a *man-made* portal to hell and thus that the whole of society is directly implicated in the creation of such places of "torment and suffering." The living created the dead; the free created their own conditions of unfreedom. However much ghost stories elide this connection by transforming difficult histories into lighthearted amusement, that they should be so pervasive and popular speaks to a cultural impulse to simultaneously remember that connection and forget it.[44] So Lindberg and Tenney come full circle, back to where they started, to spectral subjects and purgatories lost in their search for the paranormal.

Unlike the ruinscapes and monoprints of Holmesburg, *Ghost Stalkers* never expressly addresses the medical experiments conducted there, saying only of Kligman's former research hub, H Block, that it has had "its share of tragedy." The episode's focus on H Block instead centers an inexplicable death occurring well after the experimental program closed in 1974. The prisoner who died there had suddenly collapsed after walking out of the hallway, and the reason for his death remains unknown. Investigating H Block solely for this particular mystery, the cast members depart from online ghost stories of Holmesburg that do make reference to Kligman's experiments, cataloguing them as one among many scandalous events in the prison's history. The episode is haunted by this absence of Kligman's research program, featuring one of the most infamous areas of the prison, H Block, to which cast members physically and discursively return again and again without ever mentioning the main cause of its infamy. This is not to say that including its history of human experiments would have necessarily made the episode's treatment of Holmesburg any more accurate or less problematic, if online ghost stories are any indication. For if Kligman's research program could be said to attain a ghostly presence in the show, then it, too, appears only indirectly.

This absence is most conspicuous during a scene where Lindberg helps assemble the wormhole detector. While preparing the detector for his individual tour, Lindberg recounts his family's long record in law enforcement, relating his father's and his great grandfather's days in the police force. Wondering how the spirits of Holmesburg would behave toward him, Lindberg asks, "Are they going to respond to me in a negative way or are they going to . . . *respect* me?" He decides to test his query, taunting the ghosts of prisoners into activity that could be picked up by the wormhole detector. From the center of the terror dome, he bellows at them, "Time to lock down! Get *back* in your cells! . . . *I said* get back in your cells!" He reenacts guard duties, channeling constabulary power handed down through paternal succession and which he hopes would sufficiently provoke the spirits into manifesting.[45] It does not, and one can easily consider Lindberg's antics blustering and indulgent. However, they were done in the name of paranormal science—Lindberg wanted to assess the wormhole detector and thought recapitulating prison protocols would help him do that. After all, those protocols had been successful before and for a far more respected researcher, but here they surface again for much the same reason: to transform those still held inside the prison into proper objects of scientific study, test subjects. Even in its afterlife, Holmesburg remains effective, its walls now barring the escape of ghosts while also allowing access to them by inquisitive persons—these ghosts who can miraculously cross gateways between different worlds, except the one which would grant them their freedom. Tenney professes, "I think that even people who were here incarcerated and got out, it feels like even if they died somewhere else, the negative part of them gets sucked back through that portal here"—as if to say that neither death nor release from prison guarantees salvation, that hauntings merely extend in another dimension the carceral work of Holmesburg Prison in its heyday.

The Gut of Architecture

From the skeletal remains of Holmesburg Prison to the voracious ghosts that haunt it, another motif in its popular visual archive emerges: hunger. Having been deprived of state support—bodies monetized by the state—the shuttered prison inevitably devolved into the wasted space it is today. This recalls its other history of consumption and deprivation, when prisoners had deliberately gone hungry to protest their conditions of imprisonment. Self-starvation had been the most prevalent form of resistance among prisoners at Holmesburg, these hunger strikes often initiated in response to the "monotony" and deteriorating quality of meals served, as well as the increasingly cramped and violent living conditions that stemmed from overcrowding.[46] At the height of Kligman's research, the prison held over a thousand men behind bars, nearly double its capacity and providing so little in the way of care, rehabilitation, and jobs training that Kligman's experiments became prisoners' main source of income and medical attention. This led one prison chaplain to decry that the institution actively "cannibalized inmates," who survived only by "selling their blood" to scientists.[47] Though the strikes themselves were brief (lasting only a few days) and did not appear successful in their aims, their recurrence throughout Holmesburg's institutional history demonstrates how both state repression and the political action against it can share the same space and inhabit the same body.

Patrick Anderson (2010) studies subjectivation in the context of self-starvation, this being-toward-death intertwining the destruction and production of the subject through a willful staging of their disappearance. He writes of prison hunger strikes:

> The body of the hunger striker . . . asserts itself as a body, as a visceral representative for state-produced delinquency, by performing its own gradual decline, through self-consumption, to death. And so that body becomes not only the object of state punishment and torture, but simultaneously an agent imminently responsible for performing violence upon itself . . . a seizure of state power, especially the state's power to enact violence upon its subjects. (p. 123)

Appropriating the powers of the carceral space and turning consumption onto themselves, hunger strikers actively re-territorialize onto their bodies the prison's enactment of living death, embracing their potential death as a political statement against what had already been their hell on earth.[48] Hunger strikes are not unrelated to violent uprisings or more direct clashes with prison officials. Often dismissed or denigrated as "riots," outbursts of open confrontation with authority shared with self-starvation the very real risk and the very real pain of losing one's life. Still, in a space built to entomb and wilt the body away—whittled down to the most basic, rawest form of a public commodity—challenging the prison's death-dealing machinery through violence turned against the self, is life weaponized against its own conditions of possibility.

Prison scholars have used metaphors of eating and digestion to mark the exceptional place of incarceration among other built environments. Angela Davis (1998), for instance, spotlights Foucault's significant re-thinking of his own work on prisons upon his visit to Attica, New York, in 1972, a year after its historic prisoner uprising. There, Foucault saw that US prisons did not so much produce virtuous souls out of speaking subjects as discard again and again raced persons en masse. He recounted, "Attica is . . . a form of prodigious stomach, a kidney that consumes, destroys, breaks up, and then rejects, and that consumes to eliminate what it has already eliminated" (Foucault & Simon, 1991, p. 27).[49] Dylan Rodríguez (2006) similarly argues that the prison constitutes neither a destination nor an origin of raced persons but a *passage* or passageway through which racialization takes place, a "point of *massive human departure*—from civil society, the free world, and the mesh of affective social bonds and relations that produce varieties of 'human' family and community" (p. 227, emphasis his). Prisons are the underbelly, the viscera, of a body politic sustained through processes of racialization figured as acts of devouring and expelling persons from political recognition. In both these interpretations of prison space, movement loses all meaning, becoming untethered from willful action and from life itself. Inertia or a kind of death is implied in both readings, not quite a passing through but a passing away. And no doubt this vexes traditional definitions of space, place, and location, disturbing one's somatic relationship to built environments. Geographer Yi-Fu Tuan (2001) differentiates between space and place, writing that space is abstract, connoting expansiveness, unfamiliar, perhaps dangerous terrain, a venturing into the unknown and, mostly importantly, the freedom to move. Place, on the other hand, is a "pause" in this movement, a home, a familiar location, a concretization of prior experiences. But prisons are neither space nor place, instead perpetuating and propagating the "entrails of power" Saidiya Hartman (2007) observes in the built environment of the trans-Atlantic slave trade, a "usurping and consuming [of] life" (p. 114) that converted human beings into waste and that waste into capital. At the "interface between life and death," captives, or persons transformed into refuse, became material "proof that the powerful had eaten" (pp. 114–115), that modernity and capitalism depended and thrived on the accretion of eliminated human beings.

Earlier, the chapter mentioned that captive test subjects were compared to animal models in experimentation. Yet seeing the carceral system as a space for digestion suggests that they were not even that, or that they more approximated a different kind of beast. I am reminded of a 1968 testimony given by one Holmesburg prisoner regarding the movement of bodies within that space. Quite instructive, I provide an extended excerpt here:

> Prisoners confined in Philadelphia's three prisons commute from their institutions to courts by way of a prison van. The van is a truck externally resembling the sort of refrigerated delivery truck that delivers meat to food stores. The body of the truck has no windows. At the very top of the truck there is a tiny row of slots purportedly for ventilating purposes.
>
> Winter— . . . There, some 40 prisoners . . . (packed like sardines in a steel-barred can), are loaded into the van. It has only seating capacity for 15. The rest must make themselves "comfortable" as best they can. There are no handholds. There is no heat. It is freezing with an intensity so great that some prisoners relinquish their seats: The pain of frozen iron pressed against their backsides is unendurable. . . . The trip from north Philadelphia is an hour of grinding stops and bumping halts. . . . There is no light in the vehicle and the darkness is punctured by the grunts and groans.
>
> Summer—The prison van is a sweltering cauldron of red-cast iron. The packed bodies of men stink . . . the waiting becomes interminable

and finally unbearable. The prisoners scream and bang on the sides of the van but there is no relief. The time never gets shorter, sometimes it gets longer. . . .

I know, as a matter of fact, that the Interstate Commerce Commission requires that certain minimum space be provided for each individual hog shipped in commerce. Couldn't untried prisoners get the same that a pig gets?[50]

If, then, photographs, prints, and ghost stories of Holmesburg refer to an architectural body, perhaps it is not to an external skin and its associated functions of touching and feeling. Rather, they evoke an altogether different integumentary structure: the gut, the bowels, and its digestive operations.[51] Hence, to look upon Holmesburg today in all of its spectacular decrepitude is to *see what it has always been*, a place of wreckage and waste, a ruin that in an ever-expanding prison nation is paradoxically more alive than dead in its capacity for endless renewal and accumulation.

4

Bioethics and the Skin of Words

The beginning of the end of Kligman's two-decade-long research program at Holmesburg was augured by a 1968 report on sexual assault among men in Philadelphia's prison system. Authorized by the county's Court of Common Pleas and headed by chief assistant district attorney Alan J. Davis, the report, titled *Sexual Assaults on Prisoners in the Philadelphia Prison System and Sheriff's Vans* (henceforth, the *Davis Report*) detailed findings from thousands of interviews conducted by Davis and his team over a twenty-six-month period. It included testimonies from prisoners and staff across the city's three correctional facilities and uncovered an "epidemic" of sexual violence in all of them. Even its estimate of two thousand assaults for the sixty thousand captives siphoned through the prison system in a given year was deemed a very conservative figure, because survivors were generally reluctant to speak with investigators for fear of retaliation from their attackers and from prison officials who were actively suppressing such complaints. At Holmesburg, the report argued, Kligman's research program became a key contributor to the prevalence of rape and assault by placing some prisoner technicians in positions of relatively high economic power that enabled them to bribe or coerce other prisoners. Responsible for test recruitment and the disbursement of thousands of dollars of compensation, these prisoner assistants played a critical role in the bureaucratic management of experiments, becoming important gatekeepers in a program that operated much like a "separate government inside the prison system," whose autonomy was purchased through financial kickbacks, about 20 percent of prisoner wages, to the facility itself. Kligman objected to the report's conclusions, calling any associations made between his experiments and the prevalence of sexual violence "vile."[1] But despite his opposition, the report instigated formal consideration of terminating entirely Kligman's involvement with Holmesburg.

Shuttering Kligman's research program, however, took much longer to realize, only ending six years after the *Davis Report* was published and when human rights issues in medical experiments reached broadscale national scrutiny. In light of increasing media attention on research scandals like the Tuskegee syphilis study, Army and CIA tests on mind-control drugs, experimentation on children, rank-and-file soldiers, disabled persons, and the poor, growing public outcry and demand for monitoring and regulating medical scientific protocols resulted in congressional passage of the National Research Act in 1974, the same year Kligman's program was finally closed. This legislation implemented the first federal taskforce on bioethics: the National Commission for the Protection of Human Subjects of Biomedical and Behavioral Research, now commonly known for drafting the *Belmont Report* and for instituting the internal review board. Though ending before the commission began its work, Kligman's work at Holmesburg was nevertheless inseparable from wider scientific controversies propelling the commission's investigations, which thus offer a crucial historical vantage point for evaluating the moral imperatives of shutting down human experiments at Holmesburg specifically and in all US prisons more generally. At this time, Pennsylvania joined a growing number of states that had already begun phasing out all human experimentation in penitentiaries leading up to the commission's final report in 1976, at which time only seven states still held prisons with medical research programs.[2]

The commission's charter stipulated two objectives: first, to evaluate current practices and regulatory policies regarding human experimentation, and second, to generate for the US Department of Health, Education, and Welfare (DHEW) new guidelines for conducting such experiments in the future.[3] As the site of roughly 85 percent of phase 1 clinical trials conducted during this period, prisons constituted a critical locus for the commission's assignments, which included making site visits to a number of all-male prisons, carrying out interviews of prisoners and prison staff, convening public meetings to engage concerned citizens, and inviting input from medical, policy, and ethics experts as well as from advocacy groups, community leaders, and pharmaceutical industry representatives.[4] However, much like the investigators behind the *Davis Report* in Philadelphia, the commission did not question the ethical standing of imprisonment itself, taking it for granted as an acceptable or necessary institution from which they were tasked with extirpating unacceptable or unnecessary forms of violence: for Davis, sexual assault, and for the commission, abuses in medical science. In both investigative bodies, prisons instead figured as moral actors whose managerial capacities were shaped by the very subjects of their supervision.

As told by the *Davis Report*, for instance, the supposedly pernicious and uncontrollable aggression inherent in black masculinity threatened the safety of prisoners, corrupting a carceral order vulnerable to encroaching behavioral problems first cultivated outside prison walls. "Therefore," the report concluded, "although sexual assaults within a prison system may be controlled by intensive supervision and effective programming, the pathology at the root of sexual assaults will not be eliminated until the fundamental changes are made in the outside community." Displacing the source and agent of sexual terror from the prison to the prisoner, the report framed the possible closure of Kligman's program as but one remedy against the spread of sexual predation.[5] The commission's final report, on the other hand, presented a more collaborative relationship between prisoners and researchers and between researchers and prison staff, pinpointing areas where administrative accountability could be added or improved. Ultimately, the commission did not advocate for the wholesale elimination of prison medical research programs, instead highlighting the prison's own capacity to reform itself and hence prevent abuse and promote safer, more ethical scientific practice. That the commission's recommendations nonetheless failed to stem or reverse the downward trend of carceral research programs across individual states was indicative of the intractable relationship between early US biomedicine and the postwar prison, with researchers finding in the unreformed penal facility their laboratory par excellence. Nostalgic for that era in medical science, Kligman fondly remembered when he could perform his experiments with little censure or oversight: "I began to go to [Holmesburg Prison] regularly, although I had no authorization. It was years before the authorities knew that I was conducting various studies on prison volunteers. Things were simpler then. Informed consent was unheard of. No one asked me what I was doing. It was a wonderful time."[6]

Formulating the prison as laboratory, prior chapters delved into the scientific and cultural narratives it produced, depicting how it created specialized forms of knowledge and how it more recently stages a critical interrogation of the exploitative means through which that knowledge was gained. They discussed the spatial and screen-like qualities of captive skin, examining the blurred boundaries between the photograph, the photographic apparatus, and epidermal structures transformed into visual instruments and architectural features. There, skin is a concept and material reality for studying the representational capacities that captivity allowed. The current chapter elaborates on this argument through the prison laboratory's existential crisis, where incarceration's role in postwar knowledge production came under scrutiny and critique and whose de facto end came with the commission's investigation. Key to the commission's report was the prisoner's accounting of himself as a stakeholder within a community of research participants, one among many experimental volunteers behind bars. Locating and expanding informed consent were of paramount importance, mined from prisoner interviews and elucidated through the philosophical and legal terminology of ethics. Functioning primarily as testimonies of captive choice, statements made by prisoners and their subsequent interpretations by ethics experts inscribed an ambivalent subject of medical abuse, one both exploited and agential, subordinated and self-determining, a speaking and knowing subject whose narrativized woundedness provided the empirical evidence of human test subject experience and a basis for policy-driven forms of redress and prevention.

The current chapter formulates skin as this image of autonomy and self-possession whose subject is lacking in precisely that. If the skin apparatus generated scientific work while the skin of the architecture produced a memory of the prison laboratory, the textual skin of postwar American bioethics constituted a unified body or visible shape through which the captive could be discerned or recognized as something other than itself. In this sense, the commission's report is not unlike an "imagetext," which links the imaginative work of discourse with the narrative function of pictures. All media, W. J. T. Mitchell (1994) argues, is multimedia. Still, as Mitchell also argues, this is neither to collapse nor to posit a seamless connection between images and texts but to use their proximity and interplay to mark out what makes them fundamentally heterogenous to or different from each other. One must stay with rather than attempt to reconcile this unbridgeable gap, Mitchell insists, because that gap situates an important political demand, an ethical charge to discover the difference that difference makes (p. 91). And this difference is not apparent in the content of imagetexts; that is, what they show, what they say. Having less to do with meaning or making sense, the difference itself is the conflict internal to representational form. As prior chapters have shown, what skin visualizes does not necessarily coincide with the subject or the body to which skin belongs. The current chapter examines how the commission's efforts to recover informed consent through prison reform envisioned a subject of bioethics that was not demonstratively the same as the prisoner experimented upon, even as the latter's words were made to speak for and about this new entity. The vulnerable subject, it turned out, was not captive.

The current chapter does not discount instances of captive self-making in its emphasis on the forms of domination under which they occurred, but neither does it meditate on the possibility of ethics in an ethically dubious institution like the prison. Rather, it examines the ways captive stories mediated the entwined sites and practices of medical science and imprisonment, tracing the ways prisoner consent was reconstituted in a space where coercion was both banal and the rule of law, or where bodily integrity remained under constant threat or in permanent crisis. And this is something that the *Davis Report*, even with its misplaced conclusions about prison rape, could not help but show; in a "fear-charged atmosphere," ascertaining consent was impeded by the ever-present urgency of survival strategies. Thus, this is also not to conflate sexual consent with informed consent, each implicated in very different types of violence and power imbalances. Even in its criticism of Kligman's research program, the *Davis Report* emphasized only the program's influence on the prevalence of sexual violence, its creation or deepening of *financial* hierarchy within a context of extreme privation. And as this chapter will show, the commission's concerns with prison conditions emphasized only the latter's influence on prisoners' capacity to participate willingly in experiments.

But most importantly, each report's core questions about the prison, or the possibilities they each sought to determine, were structurally different. Whereas the commission asked if informed consent was possible in such a space, the *Davis Report* wondered if its own methods were adequate—not if sexual consent was possible, but if investigators were equipped to locate it in the first place. The *Davis Report* concluded that prison conditions made it so they were not, thus recognizing the role of imprisonment in shaping the progress of their research (not surprising, given the acknowledged limitations of seeking out survivors as well).[7] Missing this reflexivity in their own report, the commission's discussion of informed consent separated its meaning from its carceral surround. The current chapter's aim is therefore neither to locate or recuperate consent and ethical research inside the prison nor to disagree with the commission and prove the impossibility of consent. Rather, it seeks to bring into focus the carceral conditions giving rise to biomedicine's system of regulation and credibility that is called "bioethics."

Regulatory Formations

The commission was formed during an unprecedented rise in human experimentation occurring at both public and private institutions, and the unveiling of abuses in pharmaceutical, university, and government-sponsored medical research programs. In a context where Nazi experiments and the Nuremberg trials constituted recent memory, such issues became subjects of impassioned debate at medical conferences and in the ethics societies that grew out of them, such as the Hastings Center, the Kennedy Institute of Ethics, and the Society for Health and Human Values. Including perspectives from philosophers, theologians, and medical practitioners and researchers, these ethics societies reflected growing public concern with the possible dangers of rapid advances in biomedicine, such as genetics, brain science, organ transplantation, assisted death, and fetal research. Entrusted with assessing and recalibrating prior research guidelines to better address the problems of an increasingly commercialized, technologized, and bureaucratized medical science, the commission would create the foundational American regulatory framework for defining and protecting vulnerable experimental test subjects.[8]

From its inception to its disbanding four years later, the eleven-member commission addressed a broad array of topics—perhaps overbroad given its timeframe—including, but not limited to: the performance of existing review boards and their possible expansion to non-DHEW related research; the theoretical and practical differences between "research" and "treatment," this boundary between medicine and science muddied by clinical trials and, significantly, the latter's penchant for using human subjects belonging to society's most vulnerable or marginalized groups; the therapeutic uses of psychosurgery for the treatment of behavioral or emotional "disturbances," with lobotomy being the most common procedure used for women patients; and the role of the public in assessing ethical and legal implications of advancements made in science and medicine.[9] Such topics were taken up by a multidisciplinary team of experts, comprising mainly doctors and lawyers and including academic scholars in the fields of psychology, behavioral biology, and bioethics, all affiliated with universities, hospitals, government agencies, or non-profit advocacy organizations.[10]

Composing half of the commission's official reports, recommendations for the use of "vulnerable" test subjects spotlighted the precarity of children, prisoners, institutionalized disabled people, and the patients, often poor, of public health clinics and programs. What were once commonplace research practices became notorious displays of human rights violations or medical mistreatment, exemplifying the "hyperpublic notice" Karla Holloway (2011) observes in the myriad ways non-normative bodies are made legible in human subjects research, bioethical regulation, and public discourse. Consider the assignation of the term *vulnerable* to these test subject populations, a category very much shaped by social inequities and informed by past abuses in medicine. That term, Holloway writes, relies on a visible social identity, which "offers a text that begins to act on what we believe we understand about a body even before any interaction *other than vision* occurs. Is this body useful, or autonomous?" (p. 116, emphasis added). Historically, the question has tacitly decided which test subject populations were institutionally expedient, easily acquired and exploitable for the simple reason that they were socially neglected and/or legally unprotected. Thus, hypervisible markers like race, gender, and ability preceded and informed test subject selection, even if implicitly only receiving formal recognition as abuse through the institutional recognition of "vulnerable" test subject groups. But this shifting image of the vulnerable test subject from one of utility to one of harm also occurred in the popular arena, where injury against marginalized peoples hold "some perverse appeal to the public's consumption" (p. 113). In each optic—scientific opportunism, legally recognized harm, spectacle of medical abuse—the vulnerable subject constituted a public text through which notions of autonomy and private personhood could be reworked and expanded.

To complicate notions of privacy and vulnerability, Holloway emphasizes the force of vision in legal and popular discourse about medical science. In her writing, social identities are hieroglyphic scripts that picture bodies and what can and cannot be done with them, conceptualizing the fundamentally imagistic quality of the language of bioethics.[11] By focusing on prisoners, this chapter does not aim to conflate bioethical issues concerning every vulnerable group with those specific to incarcerated subjects, thereby undermining the specificities of the latter as well. What this focus allows is an examination of early bioethics as a nexus of carceral regulations and their related ideological refusals, principally of abolition (a topic further discussed in the next chapter). Quite revealing of their procedures, the commission deployed the same investigative plan originally developed for prisoners in their subsequent analyses of other vulnerable subject groups. This methodological move speaks jointly to the unparalleled use of prisoners in clinical trials nationwide, the ubiquitous scenes of confinement and deprivation experienced across vulnerable test subject populations, and the conceptual uses of incarcerated subjects within a national dialogue about biomedical risks, harms, and redress.[12] If the figure of the captive was hieroglyphic, then it spoke less to a singular condition of social exile than to a generalizable mode of state and state-sanctioned repression. As this chapter will show, this generalizability underwrote the commission's reformist stance.

Central to the commission's investigations was their decision to visit four prison sites for men—Jackson State Prison, the California Medical Facility at Vacaville, Washington State Penitentiary, and the Michigan Intensive Program at Marquette—and observe and conduct interviews. Jackson State and Vacaville were primarily concerned with medical research; Washington State and Marquette were focused on therapeutic behavioral studies. Much consideration was given to each prison's role in either maintaining or eradicating completely informed consent, which was by then considered by international governing bodies an indispensable factor of ethical research practice. For example, "voluntary consent" appeared as the first principle of the Nuremberg Code (1949), a foundational precept that issued from coalition tribunals prosecuting Nazi war criminals involved in human experiments inside concentration camps. The Nuremberg Code later provided the blueprint for the Declaration of Helsinki (1964), which remains worldwide the preeminent guiding statement for conducting research on human beings.[13]

Influenced by these global charters preceding it, the commission's formulation of informed consent echoed their emphasis on the test subject's capacity to choose freely, and this meaning of informed consent prevails in American bioethics today. Informed consent mandates the removal of forms of coercion (such as threats of harm) and of undue influence (such as inordinate recruitment incentives). Before experimentation could proceed, researchers must also provide test subjects with accessible information regarding experimental risks, goals, and protocols, to which test subjects must demonstrate adequate comprehension. Though informed consent does not require that subjects become as well versed in procedures as are trained researchers, it does seek to ensure that the subject sufficiently understands their role in research as well as the possible outcomes and dangers that come with participating in it. In these ways, informed consent operationalizes "respect for persons," one of the basic principles of bioethics in which test subjects are treated as sovereign agents whose opinions and judgments concerning themselves must be recognized and honored. When this capacity for self-determination is lost or diminished, such as when the subject inhabits conditions of limited choice like the prison, respect for persons extends legal protections either excluding the subject from experimentation or putting in place additional stipulations ensuring their clear, unambiguous understanding of research activities. Voluntary compliance, according to this principle, is possible only when these conditions of informed consent are met, allowing for relatively cognizant and uncompelled choices.

Following their tour of prison facilities, the commission had initially questioned the viability of informed consent in prison settings, troubled by what they deemed "serious deficiencies in living conditions and healthcare that generally prevail in prisons."[14] Prisoner rights advocates agreed, presenting their own findings and conclusions during the commission's public meetings. Both the Prisoners' Rights Council in Philadelphia and the ACLU's National Prison Project, for instance, called for banning prison research altogether. The latter had in the year prior filed suit against the city of Baltimore on behalf of nine prison test subjects at the Maryland House of Corrections, representing the first serious legal challenge to prison experimental programs. This viewpoint was promoted by several speakers at the 1975 National Minority Conference, jointly organized by the commission and the National Urban Coalition, which urged a moratorium on all non-therapeutic experiments conducted in prisons.[15] In addition to cataloguing those "serious deficiencies in living conditions and healthcare" that compelled prisoners to join experiments, advocates pointed to the enormous sway of sentencing and prison population management. The ACLU's National Prison Project attested:

> Over 80% of the felons who were released from prison in 1970 were released conditionally on parole or through some other form of discretionary, conditional release. Thus, *in almost every prison* in the United States, the prisoner believes that the date of his release from prison, the single most important thing in his life, is subject to the whim and caprice of the prison administration and the parole board. Pleasing the prison administration and the parole board becomes one of the most important elements of prison life.[16]

Though early release or reduced time did not generally constitute a form of compensation or reward, the ACLU discovered that it weighed significantly on judgments made by prisoners anyway, who saw research cooperation as a concrete sign of their good behavior. Commonly from impoverished backgrounds with little access to quality and continuing education, the prisoners did not understand the aims and methods of research nor the legal language of release forms they were required to sign in order to participate in experiments. Thus, for prisoners, the opportunity to curry favor with those whose power decided their destiny held enormous sway in decision-making. The prospect of freedom, even conditional freedom, greatly predisposed those behind bars to view research participation as a feasible means of getting out sooner. Finding this coercion "the inevitable product" of incarceration, the ACLU concluded that "additional regulations, procedures and safeguards *will not* be able to alter [these] factors and conditions."

Holmesburg Prison typified the injurious and often deadly carceral environments identified by prisoner advocates. Just shortly before Kligman's program was discontinued, the state's Commonwealth Court ruled the prison in violation of the US Constitution's prohibition of cruel and unusual punishment. Wrote one local newspaper, "Holmesburg is so overcrowded, understaffed, infested with rats and roaches, dirty and inadequate in medical care, food and rehabilitation programs that the constitutional rights of inmates are constantly violated" (Epstein, 1973). And soon after this ruling, the entire city of Philadelphia was held in contempt by its Court of Common Pleas for failing to rectify the appalling prison conditions for which the city had been admonished only a few years earlier (Rosenfeld, 1977). In this context, Kligman's program was seen by both prisoners and researchers as offering many provisional though not unimportant alleviations. In its last year alone, Kligman's program paid out over $100,000 to its captive test subjects, an incredible sum for any prison industry at the time. And prisoner test subjects had strong incentives to participate in Kligman's medical experiments: income from the research program far exceeded that from other prison industries, with a prisoner earning anywhere from $300 to $400 a month as test subject versus the fifteen cents per day given to those performing other prison work. Opportunities to participate in the latter, moreover, were considerably lacking compared to the constant demand for bodies in Kligman's program, which thus became for many prisoners a valuable source for raising bond and commissary funds. Lastly, the program bolstered Holmesburg's nearly nonexistent medical infrastructure by bringing in trained clinical staff and state-of-the-art equipment. Said program administrator Sol McBride, "The prison didn't even have an EKG [electrocardiogram] machine. We did that work for them. Holmesburg Prison doesn't even have an x-ray machine. We bought a brand new one even though we had no need for one. I only bought it because the prison needed it. . . . I also bought an $8,000 electroencephalogram because the prison didn't have one" (Antosh, 1974). A critical source of income and medical attention, Kligman's program was later supported by more than half of Holmesburg's prisoners, who earnestly petitioned for its continued operations when the city's Prison Board of Trustees met to shut it down (Epstein, 1973). And the program's end was abrupt. "There is no phasing out, no completing any cycles," announced the prison board chairman. "We're rid of it" (Antosh, 1974). But long after its closure, Kligman remained adamant about its value for medical science research. "My view is that shutting the prison experiments down was a big mistake," he asserted in a 2006 interview. "I'm on the medical ethics committee at Penn, and I still don't see there having been anything wrong with what we were doing" (Urbina, 2006).

Like those at Holmesburg, many of the prisoners interviewed by the commission also insisted on the voluntary nature of their participation and voiced their opposition to terminating research programs at their facilities. Prisoners at Jackson State, the commission observed, "gave many reasons for volunteering for research, including better living conditions, need for good medical evaluation, and a desire to perform a worthwhile service to others, but it was clear that their overriding motivation was the money they received for participating."[17] To local and national news outlets, these captives "were indignant when asked about proposals to halt prison research," with one prisoner commenting, "It's unfair. I have a right to do what I want with myself," and another affirming, "I wouldn't be over here [in the experiment] if I thought [that I had been coerced]." And for yet another, the prison clinic provided real refuge: "The atmosphere is great. You come in here and it's a whole other attitude."[18] When interviewed prisoners did provide objections, they were for the most part leveled not at research programs but at prison facilities, which the prisoners argued had placed too many restrictions on payment and participation. Prisoners demanded greater compensation, expanded job security in other prison work, and equitable access to safe experiments—these suggested improvements championing, at least outwardly, the affordances of research programs against the failings of prison bureaucracy.

The commission's final report reflected the more policy- or regulation-centered perspectives promulgated by most participants at its public meetings, ultimately advocating a tiered system that balanced the prisoner's safety with their so-called right to participation. The latter concern, that the prisoner was *entitled* to join experiments, held precedence over the commission's deliberations and found support among captive test subjects insisting on the continued presence of research programs inside prisons. Upholding the prisoner's right to medical testing was anterior to any issue, even that of informed consent. This implemented the bioethical principle of justice, or fairness, which mandated the "equitable distribution of burdens of research no matter how large or how small those burdens may be" (p. 7). Disproportionately represented in human experiments, assuming too many of the hazards but not enough of the rewards (i.e., access to treatments developed using their bodies), prisoners were already more likely to be enrolled in the most dangerous phases of clinical trials. However, the ethical solution to this unfair selection of test subjects, according to the commission, did not require abolishing prison research programs outright, for doing so would "deprive one class of persons [prisoners] of benefits of participation in research" (p. 6), categorically ruling them out as possible test subjects and thus reproducing the same problem of inequity that the commission sought to amend. In contrast, the solution the commission offered was to spread more evenly across society the risks and benefits of human experimentation, prescribing that research programs recruit proportionally from both incarcerated and non-incarcerated populations. They specified, moreover, that the quantity and quality of risks accepted by prisoners be commensurate to those tolerated by non-incarcerated test subjects, such that a study's real and potential harms were shouldered equally by the free and the unfree. Together, these recommendations were intended to curtail researchers' overreliance on captive populations while still preserving the latter's access to medical studies, abating the liabilities of human experimentation placed immoderately on incarcerated individuals not by dismantling carceral research programs but by redistributing and diversifying the test subject pool.

Though concerned with the fair representation of prisoners among human test subjects, the commission still did not preclude studies from recruiting solely from among the incarcerated, allowing for these exceptions only under strict guidelines related either to the nature of research questions or to the everyday operations of prison population control. In the former, experiments were obligated to center their investigations around the prison system itself, addressing the concerns of prisoners as a particular social group or of imprisonment as an institution. For social scientific studies, this meant studying the causes of crime and the effects of incarceration on society and on the imprisoned, while medical studies would concentrate their questions on improving or better understanding the health and well-being of the incarcerated. Requiring that research questions concentrate on problems internal to imprisonment, this proviso shifted the role of the prisoner from instrument to subject, to being the object of research instead of merely its tool. This installed more explicitly in scientific methods the interests of prisoners and their captors.

However, nearly all clinical trials procuring their test subjects from captive populations were pursuing questions separate from those pertaining to imprisonment. In many research programs, the prisoner was not under study, though they remained integral in the analysis of other phenomena. But even for these experiments, the commission, too, made exceptions, bounding them to a litany of institutional changes: the removal of possible parole as a recruitment incentive, the program's availability to public scrutiny, the prisoner's unimpeded communication with persons on the outside, the establishment of grievance procedures, and, most importantly, verifiably adequate living environments inside prison facilities. This last order was further broken down into a list of minimum stipulations, a catalog of reformist goals dictating that the prison, among other things, not exceed its designated capacity of prisoners, provide ample recreational and work activities, house sufficiently staffed and equipped medical and mental health services, and distribute personal care items to prisoners on a regular basis—all of this to substantially reduce the coercive conditions of participation. So even if they could demonstrate the scientific need of their research and show a compelling reason for testing captive populations, experimental protocols not designed to specifically help or understand the captive subject were enjoined to locate carceral settings where the "standard of living" could guarantee a "high degree of voluntariness" on the part of the prisoner and of "openness" on the part of involved institutions. These requirements were to be enforced by independent review boards and relevant federal agencies, which would hold oversight of all prison experiments, researchers, and facilities.[19]

For their final recommendation, the commission instructed that any experimental program still in place but that did not meet their specifications be completely phased out one year following the publication of their final report. And widespread closure of existing prison research programs was something the commission found highly likely, publicly stating at the end of their deliberations, "[We are] not aware of *any* prison that could now meet our standards . . . [which] will in effect end state programs too."[20] Quite telling of the generally "unjust and inhumane" practice of imprisonment, this statement is all the more remarkable given that the commission visited only those programs operating with what had been considered best practices. Still, by narrowing researchers' access to captive populations and tethering the instrumental uses of prisoners with institutional reform, the commission systemically recodified prison research programs as privileges depending on adherence to commission recommendations.

"Skin of Words"

As the commission had predicted, its final report did not resuscitate research programs already disappearing in prisons across the country, even with the exceptions they had allowed in cases of prison reform. Most research programs were uninterested in studying prisoners, and penal facilities with such programs were generally uninterested in seriously implementing the kinds of changes that the commission imagined would engender conditions of informed consent. In fact, even prior to the commission's report, the Federal Bureau of Prisons helped spur the end of carceral research programs by phasing out biomedical and non-therapeutic experiments conducted inside federal compounds, in addition to discontinuing funding for any state facility that housed federal prisoners alongside such experiments. The bureau's decision was influenced by the oppositional stance against experimental programs taken up by the American Correctional Association, which also followed in the wake of growing public outcry about medical abuses inside prisons.[21] Its actions dismayed the commission, which found the bureau's decisions overly hasty and admonished it for not having awaited the conclusion of their own investigation.[22] In the commission's view, their final report delivered practical ways of averting biomedical abuse while preserving the prisoner's access to participatory benefits, thereby maintaining possibilities for the recruitment of captive test subjects that the bureau had now barred in full. But unlike the commission, which sought to restore informed consent in prisons, the bureau did not believe such a thing at all plausible, because of, in their own frank estimation, the prison's "inherent coercive factors."[23] Incarceration was *supposed to be* coercive—it was designed to be that way—its goals and were therefore anathema to those of bioethics. With this statement, the bureau not only flatly signaled from the outset that prison reform would not be forthcoming. It also pointedly articulated the accepted logic of captivity—that, simply put, whatever made for informed consent was categorically opposed to what US prisons avowedly existed to do. And this was hardly a controversial position held by the bureau, which had simply and unabashedly acknowledged the basic structure of the institution they oversaw.

But rather than reading the commission's recommendations as having overlooked or missed this point entirely, the coincidence between the real consequences of their final report and the accomplished aims of the bureau (i.e., the termination of most research programs) suggests a more symmetrical relationship between their seemingly disparate intentions. The bureau and the commission plainly disagreed on the place of medical science in prisons, expressing contrasting views about whether human experimentation should exist there at all. Yet for both bodies of oversight, the place of imprisonment in civil society and moral law went without saying; there would be no phasing out of the prison itself. On the need for keeping human beings behind bars, the National Commission for the Protection of Human Subjects of Biomedical and Behavioral Research and the Bureau of Prisons were in silent agreement. The prison, Angela Davis (2003) shows, had by then already constituted a natural, inevitable, or unquestioned part of postwar American life. So even as they were at odds with the bureau's wholesale rejection of medical experiments in prisons, the commission took the prison itself for granted and thus shared with the bureau an "assumptive logic" that film and cultural theorist Frank Wilderson (2010) defines as the ideological limits of discourse.[24] The ethical consideration of prison research programs foreclosed a similar consideration of imprisonment in general, since it was informed by the same organizing principle and began with the same premise, as was the bureau's blanket repudiation of captive test subject experimentation. Working within a framework that preserved the prison, the commission sought only to troubleshoot those activities impeding ethical research practice, singling them out as sources of abuse and therefore subject to bioethical intervention: solitary confinement, intimidation, physical violence, lack of care and opportunity, deprivation of basic necessities, and so on. In short, the commission remade these "coercive factors" into contingencies or rectifiable processes that, in contrast, the bureau had affirmed as the prison's inherent or necessary purpose. Incarceration meant no more than these coercive factors—and certainly not less, as the bureau would have it. But rather than countering this professed truth of imprisonment, the commission's reformist language re-described it and its elementary violence as cases of excess to be resolved or reined in.

By the time the commission was formed, both imprisonment and research experimentation in carceral settings were long established, and so the latter's transition from being routine research practice to becoming spectacle of medical abuse cannot be overstated in the commission's workings, which can be seen as a response to spectacle—they were to define and quell the crisis. On the face of it, the commission's policy verbiage appears image-less, its impassively formal legalese the kind of technical language that seemingly represses visuality. But it can also be understood as reinstating an "image of common sense," which Kara Keeling (2007) calls the habituated "perceptual schemas" (like "rights") that cement dominant organizations of subjecthood. Interviews, site visits, and final recommendations formed a collective scenario for envisioning or writing into being these captive subjects able to speak and actualize their agency, minimizing the violence of imprisonment and obfuscating the procedural nature of their terror. The commission's preponderant emphasis on unearthing *instances* of unethical research conduct disguised the banality of abuse or of abuse itself as the law and not the effect of imprisonment.

Earlier, I called the commission's report an imagetext or, specifically, the content of one, representing formal safeguards within the prison laboratory and the result of which visualized vulnerability and injury removed from the existential conditions of captivity. Like skin wrapped around vulnerable flesh, the report shaped a protective barrier also giving visible form to its subject, though a subject split from the undisguised violence of imprisonment. In this sense, it also echoed the "skin of words" or "body of text" that midcentury French psychoanalyst Didier Anzieu (1985) observed in the imaginary construction of the self in response to injury. The skin of words, Anzieu argued, is a "pictogrammic" discourse whose subject appears whole, stable, and undivided and who invites self-identification.[25] Though reacting to or emerging against physical harm, against "a wound to the real skin" undermining skin's function as physical vessel or division between inside and outside, the ultimate object of the skin of words is nevertheless symbolic: an image of the undamaged self. Thus, the skin of words ultimately divulges the irreconcilable gap between the materiality of injury and the restorative potential of representation.[26] Stripped of social relations, warehoused, brutalized, neglected, medically exploited, and besieged by the empty, dead time of prison existence, the captive condition epitomizes the trauma of lost boundaries on a mass scale.[27] Against this prolonged de-individuation, the commission's report, like a skin of words, is less revealing of the prison's naked violence of homogenization and dislocation and more so the imaginary relationship it created between the captive test subject and the vulnerable subject of medical research, at pains to particularize and correct a paradigmatic violence that was by its nature beyond correction and that had in fact announced itself as such.

Prison conditions did perturb the commission throughout their investigation, sensitive to the opinions of a global scientific community horrified by the continuance of US captive test subject research after World War II. Apart from the United States, no other European power had maintained such programs following the revelation during the Nuremberg trials of Nazi medical experiments on captive populations. Noting their interactions with scientists in European countries, one commission member mentioned:

> There was not a single person who did not have the same response: "Oh, that's right. You still permit research in prisons." And their perception of it was one of total disbelief that we would have a Commission . . . study whether or not this should be in fact continued. It was a very dramatic thing, brought home to me, that their perception of what we are doing over here in this particular instance is overwhelmingly wrong.[28]

The commission wrestled with the integrity of their own mission, troubled by the moral implications of even contemplating the possibility of captive test subject experimentation in the recent aftermath of Nazi science. But in ultimately enfolding the prisoner within human rights discourse, extending notions of consent and beneficence to subjects who were by definition dispossessed of them, the commission's final recommendations deferred the ethical problem intrinsic to their work by making the captive more amenable to the solution-oriented, rights-based language of bioethics. If this exhibited a profound misrecognition of prison violence, it was a misrecognition that was immensely productive, successfully enrolling US prisoners within the ranks of vulnerable test subjects, a biomedical research category of protected individuals among which the figure of the captive remains to this day. In her analysis of US internment and prison camps, Naomi Paik (2016) makes clear the constitutive role of captivity in a US political culture based on rights. Rather than strictly defining separate legal statuses, rights and rightlessness are mutually determined, such that the expanding global human rights discourse in the aftermath of the Second World War would also, in the United States, retrench its prison regime in new ways. Modern American bioethics were emblematic of this connection, where liberal principles of human rights discourse—consent, fairness, and beneficence—could coexist in the same space and in the same body alongside their nullified forms—coercion, injustice, harm. The presence and absence of rights were indissociable, both issuing from the captive's representational and material availability to experts, either scientists or their regulators. As Anzieu's skin of words illustrates, injury structures the promises of representation, the image's allure of wholeness.

This was made pointedly clear by a prisoner interviewed at Jackson State, whose candid responses to the commission's questions resulted in punitive actions taken against him and his family.[29] In a letter later smuggled out to the commission, the prisoner, John, entreats them to protect him from retaliation by prison administrators.[30] He wrote:

> At that time, you called me out for an interview. At that time, I would not say much, and one of your people ask me how come. I told this man that most of the people that you had called out were stool pigeons and nobody would say anything as long as they were there, and if you wanted to know how we felt we would have to talk to you without no cops or prison official or stool pigeons. . . . Well I know now that I should not have talked to you people. We all told you what would happen to us if the prison officials found out.

It is unknown how officials or their informants, the "stool pigeons," learned of John's statements, but he related that in the year since the commission's visit to Jackson State, those officials had subsequently removed him from his work manufacturing license plates, placed him in "the hole," or solitary confinement, on dubious accusations of theft and assault, and moved him to Marquette Prison, where, now five hundred miles from his family in Detroit, he remained in solitary confinement. Prior to the present correspondence, John speculated that as many as ten other letters from him had been intercepted and destroyed by staff, first at Jackson State and then at Marquette. As in those undelivered letters, John now demanded that the commission make good on their promise of protection:

> They say that when I learn to keep my mouth shut, they will let me out . . . . You people said that you would make sure nothing like this would happen. I have not said many of the things that have been done to me as I know you would not believe it. I hope you will do what you said you would. If you don't, then there is nothing I can do.

This letter was delivered alongside two others written by John's brother and mother, each imploring the commission to intercede on John's behalf. They detailed harassment from guards during prison visits, as well as the difficulty of reaching John after his transfer. Both letters described the "lies and fake stories" they were told about John's treatment—that his privileges would be restored and he would not be transferred—and emphasized the immense stakes of the commission's decision to act, writing that similar injustices against other prisoners and their families would be halted or prevented should the commission aid in John's plight.

The commission handled these appeals with much skepticism. Responding to John's family a month later, they maintained that "whatever [had] happened to [John] did not happen as a result of his talking with [the commission] during [their] visit or as a result of his participation in research." Having initiated an investigation of prison records following John's letter, the commission explained unequivocally that they saw "no reason or justification for intervention," attributing the source of all "disciplinary" actions taken against John to his own "difficulties" with or violations of prison rules.

I mention John's letter not to compare or determine the validity of conflicting prisoner accounts of experimental programs but to highlight the formal workings of captive testimony in a bioethics discourse that professed its defense of the safety and well-being of captive research subjects. Participating in the commission's investigation via his testimony, John was adamant about the certainty of retaliatory actions that would be taken against him and other prisoners who would oppose research programs at Jackson State. Before and after his testimony, he asserted the obligations the commission held toward their interviewees, obligations that would protect the latter from adverse consequences of their participation. But faulting John for his own struggles, the commission's overreliance on prison-official accounts and their abandonment of John's numerous, desperate appeals, demonstrated a stunning abdication of the lofty principles of bioethics they themselves propounded, the investigators' all-important responsibilities to those they recruit in their studies, in this case, the commission's own. Much like the researchers they sought to regulate, the commission created new knowledge by entering the prison, extracting information from prisoners, and building interventions that may or may not have directly addressed the interests of those held behind bars. As John's forsaken letters revealed, prisoners taking part in the commission's investigation, too, assumed the burdens of knowledge production, furnishing its raw materials in the form of testimony under the watchful eye of authorities, while in the end not benefiting from the knowledge produced, their disposability merely converted from the protocols of human experimentation to the qualified prohibitions of bioethics—the prisoner a model not of disease or medical intervention but of the vulnerable test subject.

Reconfiguring experimentation into a vehicle for prison reform whereby the captive could become a more empowered test subject and the prison a more respectable place for such tests to occur, the commission was concerned, too, with building a proper laboratory inside the penal facility, redoubling the panoptic power previously exchanged between scientists and prison officials. Rather than safeguarding John and other prisoners at Jackson State, for example, the commission's investigation heightened the precarity of their already precarious lives by bringing them under greater prison surveillance and control. Quickly dismissed by those they were addressed to, John's efforts to bring his experiences into the open later consigned him deeper into the prison system. This is not to belabor the commission's aims and effects as misguided but well-meaning compromises or as examples of failed witnessing. Nor is it to say that the commission had relinquished their own moral project, purging consent and justice of their radical potential or political meaning. Assimilated into rights discourse, imprisoned test subjects embodied what Saidiya Hartman (1997) calls the "burdened individuality of freedom," a captive subject at once "liberated and encumbered, sovereign and dominated" (p. 117). Like a skin of words, the commission's final report did not seek to eliminate the injuries of imprisonment but to enlist human experimentation in making more bearable the pain of captivity: ethical research practice through prison reform. This was a bioethics for the greater practice of prison power, an ethics for which the body and the *bios*, or *life itself*, constituted the field of reference and point of consolidation.[31]

Ethics in Captivity

In the 1970s, the captive test subject did not become a vulnerable category in isolation. This reconceptualization was achieved through analogy, where regulators compared the prisoner to other test subject populations, thus rethinking the captive condition as one of constricted rather than nullified choice, situated along a spectrum of freedoms. Envisioned as a particular kind of research volunteer among other volunteers, captive test subjects retained their capacity to choose, composing a piece of a "whole pie" of volunteers that FDA Bureau of Drugs director J. Richard Court called human test subject populations in his address to a 1975 congressional meeting on human subject experimentation. That meeting was one among a series led by Senator Edward Kennedy and resulted in the introduction of a new bill, HR3603, eliminating all medical research conducted in federal prisons, proposed legislation opposed by the pharmaceutical industry and by DHEW, with one official noting that the bill "would prohibit certain important research activities."[32] Exhorting that prison research programs be considered in this light—a mere slice of a demographic pie—Director Court argued, "All human research is more or less coercive. The general problem can't be solved just by getting rid of prison research."[33] The difference between test subjects, he insisted, was "the difference [of] only one degree," a framework allowing him to equate the prisoner to "housewives" desirous of supplemental income and to terminal cancer patients desperate for potentially life-saving but untested treatments. All, Court maintained, had urgent reasons for participating in experiments. Court's casual comparison left his listeners aghast, with one House representative pressing, "Do you mean to tell me that you actually don't see a difference?" He did not, drawing not from any visible similarities between prisoners, housewives, and cancer patients but from internal or invisible motivations that could be verbalized. He then counseled that Congress await the commission's report to be issued a year after the briefing.

Although the report did not use Court's outrageous language, it did recapitulate his analogy. It, too, likened prisoners to other subject pools but shifted its discussion away from individual motives to issues of inclusivity and representation, inaugurating the "biomulticulturalism" that Steven Epstein (2007) observes in contemporary human subjects research in medicine: prisoners should have the same right to participate in experiments as do other civil subjects, because excluding them from participation is an injustice in itself.[34] This resemblance made between prisoners and other potential test subjects was not always explicit. The analogy was also accomplished through a pros-and-cons approach that framed captive participation in relativist terms, as seen, for example, in the commission's summary of prisoner testimonies from Jackson State:

> Overall impressions from this experience were that prisoner-participants valued the research opportunity. In general, they felt that they were free to volunteer for or withdraw from the program at will and were given adequate information about research protocols. Nonparticipants expressed various reasons why research was not for them, but did not object to its being available for others. (p. 35)

Connecting test subjects in captivity to those among the free occurred not only in the stated similarities between them (their possession of human rights) but also in the ways prisoners were emplotted into a generalizable discourse of positive versus negative attitudes. Within this framework, the legibility of captive stories and demands rested on weighing the good with the bad of captive test subject experimentation, a rhetorical strategy which determined that, in the final tally, the latter could constitute a favorable addition to prison existence, an "opportunity" that prisoners were generally "free" to take up or forego. Whereas prisoner advocates and abolitionists emphasized the desperate conditions compelling prisoners to join research programs, the final report saw such programs to be one of the better aspects of prison life granting captive subjects recourse for alleviating their everyday suffering and humiliations. This argument reversed the causal relationship found between duress and experimentation, where coercive conditions impinged on informed consent. Reformulated as opportunity and choice, participation in experiments now constituted a form of empowerment: coercion did not prompt prisoners to join experiments; joining experiments enabled prisoners to actively mitigate the coercive nature of imprisonment. In earlier deliberations, the commission had also pondered if "coercion" really did constitute a proper designation for the ethical dilemma facing informed consent in penal facilities. Putting forward "constraint" as perhaps a "more applicable" characterization, they noted, "Coercion implies being pushed into something; constraint refers to the *range of available options.*"[35]

Courtney Baker (2015) argues that the notion of human or humanity is "less of a condition than it is an idea" and that the recognition of humanity, via rights or protest, is always a "matter of perception." "The terrain of contestation," she writes, "is not reality but rather the image" (p. 9), which in the context of postwar America was predominantly a liberal, universal humanity "enamored with the vision of an equally accessed American Dream" and bolstered by a "national self-regard as 'the land of the free'" (p. 101). The commission's visit to penal facilities and their subsequent deliberations were for all intents and purposes a humanitarian consideration of all stakeholders involved, and yet the image of the vulnerable test subject in their report not only mystified the violence inherent to caging human beings, transforming it into a set of identifiably horrible but adjustable circumstances—differences of degree rather than of kind, rhetorically enabling captive assimilation into the rights discourse of the free. It also *expanded* the scopic violence of the prison's manufactured homogeneity, an extractive relationship visualizing a protected captive personhood in the limelight of human rights discourse, while prisoners like John remained secreted away, exiled from the political recognition for which their stories were first mobilized and appropriated. Adopting a reformist approach to prison violence, the report displaced accountability elsewhere, in exploitative research protocols and especially grievous prison conditions. The latter formed the rationalization of prison violence, explanations that sought sources of and solutions to that violence other than the nature of imprisonment itself, an irrevocable violence against which the commission, to borrow their own words, had "no reason or justification for intervention."

But as Baker emphasized, images of humanity often constitute a terrain of contestation, "a site of battle" between competing ideologies of who gets to have rights or of what those rights even mean (p. 102). The vulnerable test subject in the commission's report found critique not only among prisoner advocates like the ACLU. As John's letters demonstrated, captive testimony was not uniform but conflictual and sometimes radically arbitrary, pushing back against the commission's knowledge project and thus the representational efficacy of the vulnerable subject category. This was apparent in the public testimony of former Philadelphia prisoner Allen Lawson, who categorically opposed research programs in prisons. During his testimony, commission members pressed Lawson to explain why some prisoners would rather risk hearing impairment working inside a debilitatingly loud prison stamping factory rather than enter the prison's resident research program. An excerpt of the testimony is as follows:

[COMMISSION MEMBER] ROBERT COOKE: But is that [stamping factory] policed any better? We saw some pretty noisy operations going on that nobody was questioning.

LAWSON: Probably not in the prison. That is why I see such great potential for abuse of a person in prison.

COOKE: I don't mean to argue with you, but you are saying it would be okay to offer that job to the prisoners even though it is noisy as hell in that stamping plant, but it is not okay to offer something which is under much more scrutiny, quite honestly. There are very few cases of deafness from biomedical research but a substantial number from excess noise, but somehow the prisoners don't object to that, but they do object to biomedical. *I think it is kind of mystical.*

LAWSON: I don't particularly see it as mystical. I think I would take a job where there is a lot of noise rather than subject myself to the pharmaceutical company. *I may not be able to tell you why.* (Institute of Society, Ethics, and the Life Sciences, 1977; emphasis added.)

The director of the Prisoners' Rights Council, Philadelphia, at the time of his testimony, Lawson echoed arguments presented by other advocacy groups, emphasizing the "great potential for abuse" inherent to any industry operating inside a prison, be it scientific research or stamping factory. But because he was unwilling or unable to clearly distinguish the stamping factory as the better or safer option for prisoners, departing from the commission's perspicuous compare-and-contrast approach, Lawson's testimony could only be considered "mystical." (Notice, for instance, the juxtaposition of the stamping factory and biomedical research, which the interviewer argued to be "quite honestly" more regulated or "under much more scrutiny," an astonishing comment given that the commission was formed precisely for reasons of greater oversight.) And Lawson's answers *were* mystical, because of their refusal to counterpose the troubles and merits of research against those of the stamping factory, because of their total rejection of a schema resolving the obscurity of captive "choice" into the unambiguous terms of cost-benefit analysis. In contrast, Lawson's testimony revealed the impossibility of making sense of captive participation, of rationalizing why, in prison, one form of abuse would be more preferable to another or one kind of violation more tolerable. For Lawson, desperation could not be reasoned with, at least not in the way informed consent demanded.

Suspending the search for meaning behind prisoners' choices, Lawson's statements challenged the analogical structure of the commission's questioning, a line of inquiry disentangling captive choice from the singular brutality of incarceration and thus suffusing it with meanings that it lacked, meanings that could be *understood* by the free. Significantly, as Lawson also emphasized, this made captive choice markedly transparent. These choices may be inexplicable in the commission's terms, but there was nothing particularly cryptic about them in their own context—no secret, no hidden intentions, no mysterious logic to be brought to light. On captive intentions, Lawson was similarly forthright:

[COMMISSION STAFF] JOHN IRWIN: If you could now go and talk to
these prisoners in various places, would they come down in favor of
or against experimentation?

LAWSON: I think that if you approached the population and asked them
if they would like to have testing, you would still get two divided
schools of thought.

Referring to their split opinions on research programs, Lawson's reply
not only prefigured how prisoner testimonies would be summarized in
the committee's final report, able to articulate back to his interviewers
the comparative format of their questions and assumptions. Lawson
underscored the inevitability of conflict inside prisons, showing that the
"two divided schools of thought" were added tensions that research programs brought to the prison space:

LAWSON: Well, the people who were opposed to the testing were
opposed mostly on suspicions of genocide. Those who were willing
to let the tests come in were those who put their economic needs
first. Naturally, there was great hostility between them. You would
blame this guy over here for his philosophical or other reasons, and
that creates hostility in the prisons.

While some prisoners sought to capitalize on human experimentation, drawn in by its monetary rewards, others rebuked it as an instrument of mass death, dredging up the commission's unsettled reckoning
with Nazi science. But Lawson's testimony, constrained as it was to focus
on disagreements between prisoners, showed survival to be the common denominator behind every captive encounter with prison authority. Whether in the form of research participation or of abstention from
it, captive survival was varied, inconsistent, contradictory. Enduring
imprisonment meant making choices that were as endlessly heterogenous as the violence threatening it, pushing it to its multiply conflicting forms. Baffled by such choices, Lawson's interviewer betrayed his
insistence on a rationale apart from or deeper than the captive's plain
desire to stay alive and diminish their suffering, which is, Paik writes,
"the foundation for the captives' knowledge" that brings us to the edges
of understanding (p. 15). What his interviewer thought mystical, Lawson

simply showed to be the pervasiveness of violence in the prison system, a violence already out there in the open, baldly self-evident in the quotidian terrors of imprisonment rather than concealed in the private intentions of its captive targets. As illustrated by Lawson's testimony, there was nothing enigmatic about captive choices, because they were self-explanatory, indexing a violence whose nakedness or ordinariness defied further elucidation.

For Baker, a "spirit of ethics" or "humane insight" inheres in the difference between image and body, between object and subject of spectatorship, and between the myth of universalism in "the gaze" and the situatedness of "the look." Thus, humane insight attunes to the distance between self and other rather than closing it. For Mitchell, ethics arises from the unbridgeable gap between image and text that the imagetext does not paper over but brings into relief. In his writing, the imagetext activates that gap instead of figuring it as problem of representation. And though Anzieu did not explicitly refer to an ethics in the skin of words, its central function is to highlight an essential difference that is lost when skin is wounded: the difference between inside and outside or between subject and environment that installs both the self and its other.[36] The skin of words, in short, restores borders and binaries. Evident in each of these conceptualizations of difference is difference's ontological weight, which, though never absolute or self-generating, problematizes the epistemological closures performed by analogical or universalist thinking. The affirmation of alterity, or the recognition of subjective boundaries, brings forth an ethic that is relational or intersubjective because it demands proximity, contact, and connection with what is not self-same. And as Lawson's testimony demonstrated, that contact need not lead to a full understanding of difference and that a full understanding of difference need not hinder political action. Like those of other advocates, his testimony maintained the prison's exceptional place in civil society, remaining steadfast in its call to end human experiments in prisons even as its arguments perplexed the commission. In contrast, within the commission's inclusionary framework, prisoner testimony functioned to bring the captive back into the world of civil subjects, marshaled into extending notions of informed consent to prisoners, who were by definition not simply outside of their discourse but also the very embodiment of their limits. Erasing the pivotal separation between the free and unfree by conceiving both as rights-bearing subjects, or as subjects without others, the commission's investigation commandeered the living dialogue of prisoner testimonies to generate a subject of bioethics separate from the one speaking it into existence, the one giving it its voice and appearance.

While scouring the archive for correspondences sent to the commission, I came across a surprising imagetext. Another prisoner at Jackson State had requested an audience with the commission, his letter accusing researchers and prison authorities of secret experiments performed on unwilling prisoners where he was held. Active in reform work, the prisoner, Forest, was wary that guards would obstruct his prospective meeting with the commission by presenting the latter with someone else, another prisoner more sympathetic to the research program at Jackson. He claimed this deception to be common practice, writing, "I request that some member of your group personally contact me and obtain fingerprints to be certain of my identity. That might sound a bit weird but prison officials have already done illegal things to prevent me from getting the torture stopped. Thus, I give credence to the stories that some actions were hushed up by showing investigators some person other than the one they sought." To verify his identity, Forest left his fingerprints along the side and bottom edges of his letter, each digit evenly spaced and carefully numbered and classified as left or right hand. A literal imagetext, combining inscription with imprint, the letter served as visual authentication of someone made fungible by prison mechanisms of suppression, appropriating a routine technology of surveillance and identification to temporarily break Forest out from the prison's invisibleizing procedures. Willingly given, the fingerprints evoke the functions of touch, a way for the commission to come closer to or get a hold of an individual made elusive or unseeable, crowded away by other captive faces. For Forest, it was not enough that the commission read his letter; it was not enough even for them to believe its contents. Demanding presence in the absence of his actual person, Forest implored that the commission instead *use* his letter as a tool for seeing and *for being seen* by him in the flesh, making of his letter a location device against the prison's violence of mass accumulation and equivalence, an imagetext not for representation but for direct contact.

Forest ended his letter by appealing to the invisible, asking the commission to "visualize" the covert practices of prison authorities who had also hindered him from collecting proof of what he was describing: "Likewise prison officials prevent me from obtaining a camera and tape recorder to preserve evidence of the work of the 'midnight doctors,' however other prisoners are allowed to have such devices for prison assignments. (Like making newspaper articles telling how great the officials are . . .)." Insisting on the impossibility of certainty in a prison so secretive about its research program, Forest cautioned the commission to refrain from "superficial" appraisals during their visit to the prison, to look deeper than what was presented to them by prison officials, for what and who the commission observed could "actually" be other than what or who they were "seeking."

It is uncertain whether the commission ever responded to Forest's letter or met with him in person, but his allegations raised the same concerns that the investigators held about prison research programs: the problem of consent. "I am the target for and recipient of harassment and torturing with drugs," Forest wrote. "I have not agreed for any testing etc. however prison officers use me anyway." He described being forcefully injected with drugs for behavior modification and having his appeals to be examined by doctors unaffiliated with the prison ignored. And just as the prison prevented him from documenting such abuses against himself and other prisoners, the commission's probe, Forest warned, would be similarly hampered, diverted to officially accepted narratives. The commission's work was in danger, jeopardized by dishonest researchers and prison administrators. Portraying the futility of informed consent in prisons, the letter's attempt to disrupt the various ways prison authorities had muzzled Forest's truth claims, was not only an insistence of individuality made visible through fingerprints. It was also an effort to influence the course of the commission's investigation rather than be defined by statements sanctioned by prison authorities. Forest believed that he and the commission shared the same undertaking toward truth, the same knowledge project.

Revealing the treacherous context of the commission's work, Forest's letter brings into focus the systematic undermining of data gathering. His letter, in other words, raised the question of validity: How sound are the commission's observations and assessments given the likelihood of false or incomplete data provided them? Validity is elementary to any scientific investigation, an injunction that the study actually measures what it set out to measure. But Forest's letter suggests otherwise for the commission's own research, depicting the various ways prison authorities could have undermined it. These were serious methodological problems issuing directly from prison power. In turn, they call into question the commission's final recommendations, the regulations created for the captive test subject. *They call into question the status of prisoners as a vulnerable test subject category.* What *is* the captive to the vulnerable test subject? And yet, as Forest's letter also acutely makes clear in the aftermath of the commission's report, so much research could be produced in the prison anyway, so much knowledge created in a space it had described as one of "not knowing." As if to ask, yoked to the carceral, what is knowing to liberation?

Coda

War Wounds

In 2000, nearly three hundred former Holmesburg test subjects filed suit against Kligman, the city of Philadelphia, the University of Pennsylvania, and two pharmaceutical conglomerates involved in Kligman's work, Johnson & Johnson and Dow Chemical Company. Their suit claimed that the defendants had performed dangerous experiments without the consent of test subjects, including experiments with infectious diseases, psychotropic drugs, radiation, and the highly toxic pollutant dioxin, in addition to Kligman's copious studies on skin. Stipulating $50,000 from each defendant, the suit also sought formal apologies and assurance of medical care in the aftermath of tests where subjects had been underpaid and under-informed about the nature of their participation. But echoing Kligman's longstanding disregard of such accusations, the University of Pennsylvania also maintained the appropriateness of his work, stating, "during the 1950s and 1960s, the use of willing, compensated prisoners for biomedical research was a commonly accepted practice by this nation's scientists—most of whom were associated with major universities or the federal government."[1] The suit, however, was an indictment of the normalization of prison research programs in the postwar period, showing that accepted practice does not or should not translate to ethical standing. The Federal District Court dismissed the case two years later and an appeal was similarly denied in 2007, both courts citing that the statute of limitations had passed. Judge William H. Yohn Jr. wrote in his 2007 order, "As regrettable as the testing program at Holmesburg may have been, I am compelled to conclude that plaintiff's claims at this late date, fifty years later and ten years after plaintiff admitted his knowledge of the effects on his health, are all barred by the applicable statute of limitations."[2] Significantly, that order also deemed unsubstantial the appeal's accusation of cruel and unusual punishment, since the latter can only be directed at government and state institutions. This effectively separated Kligman, the University of Pennsylvania, and implicated private corporations from the carceral institutions they had relied on or shared spaces and practices with.

Following the dismissal of the initial suit, the College of Physicians honored Kligman with a lifetime achievement award for his major contributions to dermatology. Former Holmesburg test subjects and their allies did join the celebration, marching along the entrance of the College of Physicians and unveiling their own prize for Kligman, a "Doctor Mengele Award" linking the dermatologist to the infamous SS physician who conducted human experiments at the WWII Nazi concentration camp at Auschwitz. Calling themselves the Holmesburg Survivors, the protesters presented a certificate that read, "We, the experimentation survivors present [this award] for [Kligman's] consistent, steady and unrelenting abuse of man." Publicly commemorating Kligman's contributions differently from the ways the College of Physicians and the University of Pennsylvania endeavored to remember them, the survivors were "skeletons in Penn's closet," as one local news outlet described them, out in the streets refusing to forget atrocities made in the name of knowledge. "He owes us a debt," said Anthony Edwards, a former test subject at Holmesburg now living with the health burdens of those tests, joining other demonstrators to contest Kligman's official legacy in the public sphere and among health professionals.[3] Dubbed the "Angel of Death," Josef Mengele had escaped to Argentina following the victory of the Allied Powers, evading the Nuremberg trials' charges against and subsequent convictions of twenty-three Nazi doctors of war crimes. Those trials were led by US lawyers, whose verdict resulted in the making of the Nuremberg Code. By linking Kligman's experiments with those performed by the Nazis, survivors resurrected the worries of postwar bioethicists who were acutely aware that the US had been the only country legalizing human experiments on captive populations in the aftermath of the Nuremberg trials, a scandalous affair in the eyes of US allies.

Nevertheless, those same trials categorically differentiated Nazi science from human experiments conducted in US penitentiaries. Interrogating the moral standing of US medical research, Nazi defendants cited its long tradition of exploiting captive populations, going so far as to identify the United States as an important precedent to their own work in concentration camps. To the defense, both US and Nazi doctors had displayed "biological thinking," where "the patient has become a mere object so that the human relation no longer exists and a man becomes a mere object like a mail package."[4] United States lawyers and medical experts adamantly rejected the comparison, arguing not only that US prison programs were non-coercive and met the highest standards of human research but that US prisoners were also fundamentally different from the victims of Nazi science. United States experts, as Jon M. Harkness (1996) wrote in his analysis of the proceedings, "captured the US public's most common concern about experimentation with prisoners up until the 1960s . . . the worry that vicious felons might be rewarded too greatly merely for participating in an experiment" (p. 1674). To US representatives in the trials, those incarcerated in the United States were not victims of abuse but criminals given the privilege of partaking in studies, perhaps a privilege undeserved given its renumerations. In short, US prison research programs were beneficial to test subjects in ways Nazi science could never be. Following this logic, US experiments on incarcerated individuals also helped establish the Nuremberg Code by forming the basis against which Nazi war crimes could be distinguished and redressed and by which other forms of (bio)medical abuse could then be prevented. United States captives constituted a foil for witnessing the suffering of Nazism's victims, and US prisons were made to contrast and thus accentuate the horrors of concentration camps.

However, by comparing Kligman to Mengele, the protesters at Kligman's achievement ceremony unequivocally reaffirmed this connection between state violence and science. This might be read as an attempt to be included into an international code of ethics that had from the beginning repudiated experimental abuse in US prisons. Still, it is telling that the survivors directly referenced Mengele himself rather than any of the twenty-three Nazi doctors sentenced to death and hung. Like Mengele, Kligman and his collaborators and supporting institutions "got away" with their atrocities. The survivors performed their position at the juridical limits of this founding document on international bio-ethics, a venerated script later serving as the framework for developing and elaborating more principles of research conduct in keeping with human rights. Through public demonstration, they enacted what Gail

Weiss (2013) calls "bodily imperatives," an affirmation of freedom that entails also the ability to will that freedom. "Willing one's freedom," she writes, "is not always possible if the nature of one's oppression has been concealed through a pervasive mystification" (p. 148). Performing a demystification of sorts, the survivors threw into sharp relief the extractive relationships internal to both postwar medical science and bioethical regulation, abuse and its top-down correctives that abrogated freedom. This was an unofficial demystification intentionally aimed at the failed ethics of Kligman's work but also at the failure of official ethics more generally—its reach, its efficacy, and its purpose.

The survivors' demonstration might be read not with the Nuremberg Code but with the 1951 petition of the Civil Rights Congress, *We Charge Genocide: The Crime of Government against the Negro People.* Here, a similar imperative can be seen in an appeal to international human rights immediately after WWII. Submitted to the General Assembly of the United Nations, the petition accused the US government of genocide against fifteen million black Americans. It catalogued and described rampant lynching and police violence against black people over a five-year period, connecting these acts of mass terror and intimidation to an ongoing political and economic structure that had since chattel slavery systematically endangered black life and that was both explicitly and implicitly bolstered by public officials, mass media, and what the petitioners called "monopoly capitalism." Using language from the United Nations Convention for the Prevention and Punishment for Genocide, the petitioners indicted the US for conspiracy, incitement, and intent to kill in whole or in part its black citizenry. But in ultimately dismissing the petition, the UN set aside one of its most salient points: "White supremacy at home makes for colored massacres abroad. Both reveal contempt for human life in colored skin." Identifying an affinity between the visual regimes of white mob violence with high-tech weaponry, the petition insisted, "The lyncher and the atom bomber are related." The borders drawn by state terror, the petition argued, are not geographic but racial, enacted through the skin: "If they believe abroad in the equality of the races, the solution is not for us to adopt this un-American principle of race equality but for them to adopt the white supremacy of 'true Americanism.'"

These connections identified between combat and state violence, as well as appeals made to an international code of ethics formulated in consequence to world war, indicates how the prison could wed military might with scientific practice. Following WWII, prisoners became test subjects in studies led not only by corporations and university researchers, but also by the Army and the Central Intelligence Agency. Of note were the thousands of biological warfare experiments the Army funded at the Maryland House of Corrections between 1958 to 1970, where prisoners were exposed to infectious diseases like malaria, typhoid, cholera, and others. Those studies had spurred the ACLU to file suit for $1.25 million in damages against the Army and the prison, the nation's first-ever attempt to end non-therapeutic experiments on prison populations through the courts.[5] If the prison already functioned as an apparatus of domination working on and through the captive body, military studies further demonstrated the extent to which the state can methodically infiltrate the captive subject as an object of control and knowledge making, including the development of new technologies for the subjection of others at home and abroad.

Kligman's work at Holmesburg, too, was a leading figure in this trend. Although military contracts were frequently awarded to other universities and correctional facilities involved in medical research pertinent to the armed forces, Kligman and his institution secured the very first, the greatest number, and the most sizable of these commissions.[6] His studies on skin inflammation, for example, also involved questions about the mechanism behind skin hardening, a protective response that "had both offensive and defensive military implications."[7] Literally transforming the skin into armor, Kligman's studies achieved hardening through an "intense inflammatory phase" that included the continual formation of blisters. Thuy Linh Nguyen Tu (2021) notes the racialized nature of these findings, where the skin of white test subjects could harden by "taking up or 'structurally amplifying' the cellular properties of black skin" (p. 85) already figured as "invincible barrier" (p. 72). In some of these experiments, prisoners would immerse their arms into test compounds like SLS and chlorinated phenol, an ingredient in pesticides and herbicides. Though some "contracted undesirable allergies" and even "exhibited psychotic reactions" that led to hospitalization, Kligman was able to successfully harden the skin for a year, a proof of concept for super soldiers.

Other experiments sought to weaken the enemy, testing "choking agents, nerve agents, vomiting agents, incapacitating agents and toxins."

These studies expanded the Army's arsenal of chemical warfare compounds, but the ones that garnered the most public attention were those
on mind-altering drugs like LSD. The latter research created its own infrastructure inside Holmesburg, where two aluminum trailers were installed, complete with padded cells where researchers could observe any
one of the 320 test subjects chosen to take part in the Army studies. From
1964 to 1968, Kligman would use these trailers to test the incapacitating potential of multiple psychotropic formulas, whose effects included
drowsiness, unsteadiness and clumsiness, delirium, impaired concentration, and visual illusions and hallucinations—for some test subjects, the
prison walls appeared alive and seemed to be moving and "breathing."
Now a site of combat, the psyche was brought out of the body as an observable behavioral change in response to hallucinogenic agents. One
of the Holmesburg Survivors, Anthony, recalled, "Guys on those tests
were coming out of the army trailers like zombies. I wasn't much better.
I was a mess. Everybody knew I had snapped. They all started calling me
'Outer Limits. . . .' In that first experiment, the patch test that affected my
skin and hurt so bad, I was afraid I was gonna die. But this one did such
strange stuff to my head, I really wanted to die" (as cited in Hornblum,
2007, pp. 71–73). Contrary to Kligman's reports to the Army, many prisoners like Anthony continued to experience psychological problems years
after the experiments, some actively punished with solitary confinement
for their behaviors while still at Holmesburg. Evoking the prison's capacity to consume and diminish the life of its captives, Anthony remarked
of Kligman's team, "I was rolling the dice on those doctors being human
beings, but they weren't. They were vampires. They just hurt me" (p. 75).
His participation in Kligman's experiments was not so much about consent as it was a form of gambling with his life and health, a rolling of the
dice to acquire the means of surviving prison existence. Clearly, these
tests had gone beyond the theoretical and practical limits of dermatology,
Kligman's discipline, and moved deeper, away from surfaces of the body
to the internal workings of the mind and away from issues of treatment
and safe use to weaponized psychiatry. For this reason, Allen Hornblum
dubbed Kligman's program a "Kmart of human experimentation," studies
in disregard of disciplinary commitments.

Popular criticisms of these studies on mind-incapacitating substances have raised at least two procedural concerns. First was its use of

vulnerable civilian populations in military research, evoking Nazi experiments that had only very recently been brought to light and condemned in the Nuremberg trials. Military research in prisons was, in short, science gone awry. Second were the shoddy experimental procedures and results accompanying Kligman's dabbling outside his area of expertise, with one Army clinical researcher appalled by its display of "pure gibberish . . . absolutely useless . . . list of clichés seemingly pasted together without consideration or coherence, in an attempt to provide a façade of competence and ability" (p. 133). Problematizing both the ethical and the practical significance of Army tests inside prisons, these kinds of censure condemn either their morally questionable circumstances (the use of prisoners) or their definitive value (sloppy experiments), distinguishing the context of research (the prison) from its inner mechanics (the method). As described in this book, however, if the conditions of laboratory practice rebounded on those of captivity, if the experimental and the carceral were both historically and ontologically related, then Kligman's use of prisoners was as much a formal aspect of laboratory work as were all his other technical decisions. Myriad protocols, in turn, pointed to captive skin as the epistemic conditions of scientific practice. That Kligman's analytical focus could move from the surfaces of the body to the deep workings of the mind said something about the structural indeterminacy of captive skin. In short, the prison was the method, and skin the context of research.

The Army, for instance, was cognizant of the practical affordances of prison research, having realized the inadequacy of preliminary studies that had been conducted with a limited number of civilian and military volunteers. To meet the standards of valid and reliable experimental results, Army researchers were well aware that they had to conduct human testing "on a large scale." Kligman's research program at Holmesburg was a fitting location for these studies because of its well-established application of captive test subjects and its on-site, ready-made subject pool. Compared to volunteers in uniform, prisoners constituted a much larger and more rigidly contained and monitored population. They were, in the words of Hornblum and other critics, "human guinea pigs," already widely and successfully deployed for civilian research purposes. As with civilians, prisoners made scientific sense as instruments for military research in mind-control and chemical warfare, giving Army scientists the systematicity they were looking for. Here, too, experimenting on captive subjects was neither accidental nor situationally improper but, scientifically speaking, rational. Though university psychologists and Army specialists accompanied Kligman's research team, Kligman ultimately directed the administration and evaluation of tests, his lack of expertise in psychology and psychiatry notwithstanding. The Army chose Kligman's program not for its work on skin, but for its established infrastructure within the prison facility and its illustrious affiliation with the University of Pennsylvania.

Many of these experiments did fall outside of Kligman's training, but they did not simply substitute dermatology for neurochemistry, posturing knowledge outside of Kligman's specialization. More importantly, Kligman's commissioned work on psychotropic agents demonstrated a fundamental separation internal to skin in which skin became irreducible to a specific body of knowledge, like the field of dermatology, or an empirical phenomenon of study, the dermis. Attached to captive bodies, skin became something more than its objective reality as skin.[8] Rather, it now marked the very absence of boundaries, or the various contingencies of knowledge production made possible by the captive condition. While using captive test subjects was logically consistent with the methodological demands of research, studies beyond dermatological significance reflected the epistemological flexibility of captive skin. But even within dermatology, captive skin exhibited extraordinary utility. As told in one review of Kligman's large body of work, dermatology's increasing specialization thereafter ensured that the field would never have "another Albert M. Kligman." Kligman, the authors wrote, "chose the right field at the right time," which enabled him to "pursue a wide array of subjects that were essentially black boxes as far as basic understanding."[9]

Hortense Spillers ([1987] 2003) has called the captive condition a "theft of the body" giving rise to an "unclaimed richness of possibility" or a "living laboratory," whose assortment of techniques documented in slavery archives "take on the objective description of laboratory prose— eyes, beaten out, arms, backs, skulls branded, a left jaw, a right ankle, punctured; teeth missing, as the calculated work of iron, whips, chains, knives, the canine patrol, the bullet" (p. 207). To Spillers, this wounding constitutes the founding gestures of a "politics of melanin," where skin came to materialize the undecipherable marks left by the brutalities of captivity (p. 213). However, even though this wounding itself remains undecipherable, it performs epistemological work, distinguishing the "body," which is subjectifying, limned, and situated, from the "flesh," which is atomized, turned inside out, and undifferentiated in its ripped apartness (pp. 208–215). For Spillers, flesh gives rise to bodies—in her words, it is "vestibular" to culture—bringing forth and making sense of entities other than itself.[10] If one reads captive skin as flesh, then Kligman's experiments for the Army can be understood not only as having been beyond his capabilities but as materializing this richness of possibility inhered through flesh. At Holmesburg, skin as flesh, rather than as simply dermis, contributed to a specific body of knowledge— that is, dermatology—but it also cohered the conditions of knowledge production broadly speaking, enabling disparate studies to happen and coincide, studies ranging from chemical and psychological warfare to corporate sponsored research on grooming products. It too shaped changing views of vulnerable research subjects within the US medical establishment, from the Nuremberg trials to the *Belmont Report*. This is not to conflate the interests, intentions, or outcomes of the diverse knowledge projects that took place in prisons but to illustrate the "gathering of social realities" (p. 208) that captive skin brought into focus.

In Spiller's writing, the vestibule is an important metaphor for thinking about flesh, a "pre-view" of those social realities it occasions.[11] Not to be confused with *preview*, flesh or vestibule does not signpost or constitute the beginnings of something that is to be seen more fully later. Flesh does not corporealize primordial or incomplete vision. In *pre-view*, flesh instead suggests that which issues all other vantage points and sceneries, allowing us to see and to see *from somewhere* while itself never being the object of its maneuverings—a "signifying property *plus*" (p. 203, emphasis in original). This book has situated this impossible optics in postwar prison research, calling its epistemological plenitude "skin theory." Visible and usable in enumerable ways, captive skin obtained its vestibulary function in both the realization and limits of knowledge projects, be they dermatological knowledge or labels of ethical and unethical practice. In each chapter of *Skin Theory*, the extractive, manifold uses of skin demonstrated its unrestrained malleability in the practice and production of representation. As screen, skin was a technical assemblage of looking practices resolving racial difference through skin color and perceptions of pain. As space, it materialized the intersecting geographies of imprisonment and medical science, revealing how complex relations of power and knowledge are encoded in the built environment. And as text, it imaged the vulnerable subject of early American bioethics, whose regulatory attempts to purify, correct, or reform medical science in prison both mirrored and left intact the epistemic violence of science and incarceration. Throughout *Skin Theory*, captive skin was shown to embody the visual regimes of the powerful and their forms of knowledge, an embodiment that produced, paradoxically, no body, a body without boundaries, a body *without skin*, as it were, one under constant siege from its surroundings.

Thus, like flesh, which indexes the wounds of captivity, the skin theorized and analyzed in this book is also intimately tied to injury both symbolic and material. If skin implies boundaries, intimate surfaces, and individuation, the captive condition brings it closer to a wounding, to undecipherable markings giving meanings and bodies to other signifiers besides their own. As Michelle Ann Stephens (2014) argues, "The skinned body that remains left behind by both physical captivity and cultural capture is what Spillers means by the 'flesh'" (p. 3), and Stephens, too, makes a distinction between "epidermal skin and sensational flesh as two different but linked modes of understanding, experiencing, and performing the black body" (p. 11). The Holmesburg Survivors called it a "debt." *We Charge Genocide* called it a crime against humanity. Both brought into public notice a different way of envisioning the enduring violence that engenders flesh, an unanswered accountability. In doing so, the former signaled the limits of ethical discourse and the latter its possibilities, but both their demands ultimately rejected and expunged from the frames of official narratives of atrocity and redress, their accounts returned to their originary condition as undecipherable, unassimilable text—the wound.

Writing in the postwar period, Fanon's theorization of black being also visualized it as a wounding, a fragmentation of the self between the body and the image installing and proliferating the white gaze. He called the continuity between this gaze and black being a cycle into which the latter is locked and against which it must explode to escape. In *Black Skin, White Masks*, Fanon wrote of his everyday explosions, saying, "I lose my temper, demand an explanation. . . . Nothing doing. I explode. Here are the fragments put together by another me" (p. 89). But in his later work, that explosion is no long self-directed but exteriorized, aimed at white society. The explosion in *Wretched of the Earth* (2004) is war. Destigmatizing violence as a proper response to violence, *Wretched* situates notions of "the good" in armed resistance to domination, "quite simply what hurts *them* most" (p. 14, emphasis in original). The skin in *Black Skin* reappears in *Wretched*, where Fanon defined armed struggle as "the determination to defend one's skin" (p. 89) but also an activation of the "atmospheric violence . . . rippling under [it]" (p. 31). This violence is *atmospheric* because it was put there; it is all the material and symbolic forms of white aggression internalized, or *epidermalized*, in black being. Embodying this constant besiegement, the violence rippling underneath black skin, Fanon now argued, must be returned to its proper place, not to the black self but back to the white society that first launched it. The skin in *Wretched* is thus not only an object to be protected but also the source and container of radical destructive potential.

A decolonial text, *Wretched* was widely read among black power and black liberation leaders of the 1960s and 1970s, whose war rhetoric related their freedom struggles at home with anticolonial, global uprisings against empire abroad. Dan Berger (2014) connects this revolutionary nationalism to prisoners at the time who defined captivity *as a form of nationality*, a "captive nation" imagining itself as "part of a collective force strong enough to challenge the totalizing authority of the prison" (p. 9). An embattled country within a country, the captive nation was in intimate dialogue and solidarity with the black liberation movements organizing outside of it. The notion of a captive or prison nation was also implied in the acknowledged consequences of officially limiting human experiments in prisons. Federal investigations of prison research programs and the US pharmaceuticals affected by their recommendations were well aware that the search for test subjects would expand globally should those programs end. Said one spokesperson for the Pharmaceutical Manufacturers Association, "Given this shred of regulation, the drug companies are looking at our sources, turning to foreign countries for research subjects."[12]

The Justice Department banned experimentation in federal prisons in 1976, and Congress severely restricted the practice in 1978 under Title

45 of the Code of Federal Regulations. As a consequence of these regulatory limitations on prison research, US offshoring of clinical studies has steadily increased, mainly from high- to low-income countries and many of which with colonial histories, reaching unprecedented levels at the turn of the twenty-first century, with the US leading the postindustrial nations in the number of outsourced trials. As of this writing, at least half of US clinical trials are conducted beyond national borders.[13] This is not to say that human experimentation outside of the United States did not occur prior to and during the postwar period but rather to illustrate the centrality of US imprisonment in the escalating global search for test subjects from the late twentieth to twenty-first centuries. The implications of this intensified internationalization of human experimentation for studying the confluence of medical science and captivity are at least twofold. First, the spatial histories of colonialism and imperialism are arguably at work in the changing geographies of clinical research, overdetermining formerly colonized and lower-income (areas of) nations as available pools of test subjects for globalized medical science in late capitalism. Second, it points to the other(ing/ed) space of US prisons as apart from and yet internal to the national landscape, a kind of interiorized exterior sphere of precarious citizenship that preconditioned the migration of human experimentation from the American prison-academic-industrial complex to the outsourced clinic.[14]

An exceptional space within US borders making it precisely what Angela Davis (2003) describes a "terrain of justice," the postwar prison was important ground for politicization and nation-building of a different kind, its carceral violence producing its own countervailing force. Political philosopher Joy James (2007) calls the captive condition "warfare in the American homeland." The language of armed conflict, the lingua franca of domination, is in fact central to abolitionist narratives, which recognize that the dismantling of a prison nation means, precisely, "war" (p. 9). For example, during her incarceration in Middlesex County, New Jersey, Black Panther activist Assata Shakur recorded a statement about the charges of murder and robbery leveled against her, criticizing as well the media's role in constructing and disseminating propaganda maligning her character and her involvement with the Black Liberation Army. In her statement, titled "To My People," which was radio broadcasted on the Fourth of July 1973, Shakur asserted:

I am a Black revolutionary, and, as such i am a victim of all the wrath, hatred, and slander that amerika is capable of. Like all other Black revolutionaries, amerika is trying to lynch me. . . . Every revolution in history has been accomplished by actions, although words are necessary. We must create shields that protect us and spears that penetrate our enemies. Black people must learn how to struggle by struggling. . . . We are created by our conditions. Shaped by our oppression. We are manufactured in droves in the ghetto streets, places like attica, san quentin, bedford hills, leavenworth, and sing sing. They are turning out thousands of us. . . . It is our duty to fight for our freedom. It is our duty to win. We must love each other and support each other. We have nothing to lose but our chains. (1988, pp. 52–53)

Recognizing ghettos and prisons as the ground of both oppression and radicalization, of struggle and struggling, Shakur bridged her lived experience of state violence with those "droves" and "thousands" marked out for disappearance, annihilation, and social neglect, connecting her life story to the nation's systemic brutalization of black communities. Significantly, for Shakur, insurgency constituted both a moral imperative and a form of collective care, the binding of self with subjugation under contexts of domination making it so that loving and supporting each other included a *duty* to fight for freedom and to win, or, that fighting and winning become necessary expressions of that love and support.

This message encapsulates Shakur's 1988 autobiography, which melds life story with political vision, punctuating accounts of lived experience with calls to "fight on" (p. 53). Shakur's life narrative begins with a vivid narration of her violent 1973 arrest on the New Jersey Turnpike: shot twice under police gunfire, then punched and kicked while half unconscious on the pavement and beaten again while shackled to a hospital bed for treatment of her injuries. This assault marks the very beginning of her life story; violence introduces *who* she is for the reader. The autobiography itself is non-linear, interspersing chapters on growing up in the segregated South and then in New York with chapters on her radicalization: her membership in the Black Panther Party and the Black Liberation Army, her incarceration and court hearings that she called "legal lynchings," the birth of her daughter while in captivity, and her present fugitive status in Cuba, where she remains, in her words, "a 20th century escaped slave."[15] With its irregular, discontinuous timeline, Shakur's narrative points to that circularity Fanon first theorized in *Black Skin*, in which self and self-narration issue a recurring subjugation whose end can only be met through explosion, either in the subject or in the streets.

Insurgency is also present in the prison letters written between 1964 and 1970 by Black Panther and Black Guerilla Family founder George Jackson (1994), who for ten years was held captive at Soledad Prison on patchy charges of robbery and later assassinated during an escape attempt. Politicized in prison through readings by Marx, Lenin, Trotsky, Engels, and Mao, as well as revolutionaries like Huey Newton, Bobby Seale, and Che Guevara, Jackson would produce a collection of letters, heavy with "rational rage" (p. xiv), that has since become a touchstone of radical antiestablishment literature and political philosophy, enacting what Jackson himself had promised to do even after death: "If worst comes to worst that's all right, I'll just continue the fight in hell" (p. 127). Like Shakur, Jackson saw the continuation of racial slavery in the carceral and policing formations of his time, for which he had many appellations: a "life-death cycle" (p. 70), a "concentration camp" (p. 115), "life on the installment plan" (p. 26), "the closest to being dead that one is likely to experience in this life" (p. 14). Jackson maintained that this existence was a real battlefield, a perpetual state of hostilities or "usual turmoil" (p. 104) wherein one must "expect anything, including trouble, especially trouble" (p. 112). For Jackson, the proper or just response was a countervailing force of arms, a form of domestic guerilla warfare targeting the entirety of the carceral state: "The jungle is still the jungle be it composed of trees or skyscrapers, and the law of the jungle is bite or be bitten" (p. 107). Despite his deteriorating health under the tortures of imprisonment, Jackson maintained his desire "to be in the vanguard" (ibid.).

Jackson was critical of the science of US warfare, an excellence in technology for the "breakdown of civilization. . . . This is the only thing they understand, the only thing they respect—the only thing they can do with any dexterity" (p. 67). From the tomb-like spaces of the prison, he saw that triangle of death linking together "industry, science, and strategy" that Michel Serres (1974) had called the "thanatocracy" of the postwar era. The weaponry of WWII, Serres argued, inaugurated a technoscience pervaded by the death instinct, a science that kills its own history by submitting everything to reason, including reason itself.

"The end of history, the Triumph of Reason" (p. 23), where science be-comes the logic of war.[16] Against the state's war against black and brown peoples across the nation and across the world, Jackson promulgated radicalization not to escape that war but to have a fighting chance of winning it. His entreaties and sometimes disdainful stance toward his own family's "illusions"—their "brainwashing" and complicity in "herd" mentality—were uncompromising, but they also recognized conscious-ness as another battleground between Western ideals and their opposi-tion. Indeed, Army experiments with psychochemicals literalized such statements, that the field of combat or the operations of war occur both in the mind and in the open, in the psyche and in the material world.

Militant perspectives like those of Shakur and Jackson compose a "warrior tradition" that Robin D. G. Kelley (2002) notes was but one of multiple voices and approaches in the long history of the black radical imagination, and one that did not always lead to armed resistance. How-ever, they offer a different way of approaching the post–WWII nexus of war, biomedicine, and incarceration, as well as where bioethics inter-vened and failed to intervene through its principles of autonomy, justice, and beneficence. In the last chapter, *Skin Theory* argued that bioethics constituted a textual skin, a protective envelope of legal discourse en-gendering an imagined subject of biomedical abuse. Unlike the commis-sion, however, contemporaneous black radicals like Shakur and Jackson did not seek for the captive subject state protection against systemic harms but instead took aim at the state itself as the purveyor of brutality and the source of brutality's sanction. This makes the representational and material practices of insurgent warfare compelling for its analyti-cal potential to unravel the optics of bioethics, a refusal to be symboli-cally incorporated into the language of the state.[17] Rather than taking up a project of legal representation, the language of war among radicals redeployed the structural invisibility of black pain, embracing its sense-lessness by transforming it into spectacle: armed struggle. The imagery produced out of this political philosophy, Leigh Raiford (2011) shows, was a self-conscious expropriation of spectacle, and thus the "emergence of a dialectic in which a reified blackness no longer shores up universal whiteness but asserts itself at the most inopportune moments" (p. 143).

Skin Theory has critiqued and attempted to avoid "the spectacle" in order to illustrate the banality of scientific racism, of its pervasiveness and predictable returns. The book has done so in suspicion of spectacle's mystifying effects, that it does more to simplify and cloud the systemic nature of violence than it does to reveal that violence for what it is: normal. Yet the rhetoric of black militancy redeployed the spectacle of violence in new ways, appropriating those mystifying effects to refuse the state's representational authority and thus to push at the liberatory limits of political legibility. This might be an insurgent form of bioethics, one that foregrounds the imprisoned body and its structure of demands: the eradication of a prison nation. In contradistinction to the textual skin of bioethics, militancy's focus on the wounded subject of state violence contributes not a language of protection and recompense but metaphors of war and ostentatious images of self-defense, a retaliation for harm with harm, sometimes resulting in very real armed confrontations with police and prison guards. And resistance was often met with more oppressive forms of control. Immediately following the 1971 prison uprisings in Attica and San Quentin, where Jackson was shot and killed, Philadelphia's police, prison authorities, and fire department "beefed up" security in the city's three prison facilities, especially at Holmesburg Prison.[18] Ironically named Operation Breakout, the added security combined with an "inch-by-inch" search of each facility sought to quell with force any materialization of what officials believed was the always present "potential for trouble." At Holmesburg, known "inciters" like the Black Muslims received increased scrutiny, already placed in the only maximum-security wing of the prison.

As Fanon also warned, explosion, whether subjective or real, is no promise of freedom. Neither *Black Skin* nor *Wretched* make such a claim, and in fact the latter not only delineates the multiple barriers to revolutionary action, such as lack of solidarity between classes of colonized peoples or the failure of leaders to cultivate national consciousness in favor of a romanticized, illusionary past prior to colonization. It also identifies these same barriers as the possible *consequences* of revolution—a vicious cycle. Hence, insurgency and the explosion that Fanon speaks of are all the more resistant to knowledge and representation. They are difficult to comprehend, and their futures are also difficult to image and imagine, because they are intransigent, spectacular, and demanding of action without guarantees. In *Black Skin*, Fanon writes, "In a fierce struggle I am willing to feel the shudder of death, the irreversible extinction, but also the possibility of impossibility" (p. 193). *The possibility of impossibility*—this is agonizingly hard to picture or make sense of, resolutely defiant of reasoning. But as Fanon also writes in *Wretched*, the fierce struggle that aims at nothing less than changing "the order of the world, is clearly an agenda of total disorder" (p. 2). *An agenda of total disorder*—this is an admission that knowledge, and more and more of it, does not necessarily set anyone free.

Militant abolitionist rhetoric affords a rethinking of bioethics in this way, a recognition of the antagonism between freedom and knowledge exemplified by the postwar prison laboratory. The value of spectacular violence from below, of proliferating rather than healing and representing wounds, is a reflexive accounting of the politics of knowledge production, asking not simply to what ends but for whom. This is an inhabitation of injury that aims not at fixing broken systems or protecting "vulnerable" subjects but at the total undoing of the carceral state and its attendant knowledge projects. Staying with the wound and its undecipherability thus produces its own ethics. In Fanon's words, we need still "to touch with our finger all the wounds that score our black livery" (p. 164). And if it is difficult to see this insurgency as bioethics, or if bioethics as we know it cannot imagine the world that insurgency struggles for, it is perhaps because the wound is doggedly illegible, fraught because not easily given to understanding. After all, to touch a wound is to first make it sting in pain, to watch and see it expressed as inexpressibility, a body flinching and pulling away.

ACKNOWLEDGMENTS

Right off the bat, I must thank Eric Zinner, Furqan Sayeed, Richard Felt, Martin Coleman, and others on the editorial team at NYU Press for making this book possible during the COVID-19 pandemic. I am also especially grateful to my indexer, Jess Klaassen-Wright, and to the two anonymous reviewers whose close readings of the manuscript and detailed input strengthened substantially the arguments and contributions of a manuscript whose writing was often frustrated by the pandemic. Finishing this book occurred in a health emergency of global proportions, during which time we also witnessed the January 6, 2021 attack on the US Capitol; the 2020 summer of violent police suppression of protests that called into account precisely those kinds of brutal tactics that regularly take black life; and far too many climate catastrophes, from record-breaking heatwaves and wildfires in the West to infrastructure-collapsing temperatures of well-below freezing in the South and hurricanes and floods along the East Coast. With all the ways these events have profoundly disrupted our professional and personal lives, I am deeply, deeply grateful to the people who with so much patience and care helped bring this book to fruition in such anxious and trying times.

Research for this book began during my graduate studies at the University of California, San Diego, in the inimitable Department of Communication and Science Studies Program. I would not be the scholar I am today if not for the spirited conversations, the challenging lessons, and above all, the friendships I took away from these spaces. My sincere thanks to Patrick Anderson for his unwavering belief in the project, Kelly Gates for sharp though always generous readings, Dennis Childs for exacting but wonderfully principled standards for scholarship, and Martha Lampland for sharing her consummate verve whenever my motivation and rigor flagged. Lisa Cartwright taught me how to be interdisciplinary in every good sense of the word, and I owe much of the theoretical and methodological foundation of the book to her early direction. And special thanks to Jared Sexton for years of inspiration and encouragement, of sympathetic ears and high expectations.

Getting the PhD itself marked a significant (and admittedly terrifying) career change, and I appreciate those who nudged, dared, or otherwise supported the transition: Jared again, John Yoder, Kiho Cho, Chih-Min Lin, Amy Schmitz Weiss, Barbara Mueller, Charles Goehring, Michael Roberts, David Dozier, Antoine Didienne, and Jessica Plautz. Shifting from the work of laboratory science to that of critique presented a humungous learning curve, and I hold the deepest appreciation for those who helped me keep with it, making me realize the immense value of interpretive forms of knowledge alongside demonstrable ones. Over the years, many brilliant people have been gracious sounding boards for my ideas, leading me to embrace the humanities as my new intellectual home: Mukhtara Yusuf, Jason Magabo Perez, Maisam Alomar, Vineeta Singh, Jahmese Fort Williams, Brie Iatarola, Emily York, Fiori Dalla, Sarah Klein, Anna Starshinina, Kim De Wolff, Pawan Singh, Thomas Connor, Tara-Lynne Pixley, Christina Aushauna, Yelena Gluzman, Louise Hickman, Boatema Boateng, Zeinabu Davis, Cathy Gere, Kalindi Vora, Charles Thorpe, Kimberly Juanita Brown, David Serlin, Michelle Murphy, Banu Subramaniam, Rachel Lee, Mara Mills, Deboleena Roy, Karen Barad, Lilly Irani, Elizabeth Losh, Lisa Parks, Lucy Suchman, Natasha Myers, and Elizabeth Wilson.

Now a member of the faculty at USC Annenberg, I am indebted to colleagues within and outside of my home department whose assistance and support were pivotal to my intellectual and professional progress. I would like to thank Taj Frazier, Sara Banet-Weiser, Josh Kun, François Bar, Jonathan Aronson, and Hector Amaya for their camaraderie and their efforts to support the work of junior faculty like myself. For collaboration and new directions in research, I am grateful to Allissa Richardson, Alison Trope, Henry Jenkins, Kate Levin, Nic De Dominic, and Mike Ananny. Thank you to Nitin Govil, Rhacel Parreñas, Adrian De Leon, Robin Stevens, and Ben Carrington for advice indispensable to navigating one's early academic career. I am privileged to have learned from so many brilliant graduate students at USC, whose research and teaching have informed my own, but especially that of Lauren Levitt, Pam Perrimon, Adrienne Adams, Caitlin Dobson, and Haley Hvdson.

In the years between entering graduate school and the publication of this book, the project has benefited from countless exchanges with scholars, either remote or in person at conferences and seminars, too numerous for me to hope to name here. But for their notable intellectual companionship, I am grateful to Anthony Hatch, Ruha Benjamin, Moya Bailey, Diana Leong, Sandra Harvey, James Doucet-Battle, Zakiyyah Jackson, Safiya Umoja Noble, Oona Paredes, and David Kirby.

And to my comrades James Bliss, Monika Sengul-Jones, Poyao Huang, Lihn Nguyen, Aundrey Jones, Anthony Kim, Christina Dunbar-Hester, Hye-Jin Lee, Sulafa Zidani, and Lisa Lindén, thank you for your keen minds, for your integrity, and for your deep well of patience for my dramatics.

Many, many thanks, too, to Mathew Christopher, Patricia Gómez, and María Jesús González for allowing me to use their brilliant and beautiful visual works in this book. Deriving many of its materials from the Urban Archives at Temple University and the Bioethics Research Library at Georgetown University, the project also owes much to Brenda Galloway-Wright, John Pettit, Josué Hurtado, Nat Norton, and Martina Darragh for helping find and gather the diverse objects presented herein.

The American Association of University Women, the University of California Humanities Research Institute, the University of California Institute for Research in the Arts, and a Provost Assistant Professor Fellowship (2017–2018) at the University of Southern California provided financial support for this project.

Finally, to my family: *Salamat ta inaramid nak nga* weird. *Nagsurat ti libro.*

NOTES

1 Statistics acquired from *Philadelphia Inquirer* articles (1969, June 3; 1969, October 21; 1971b, September 7) citing prison officials.

2 See James B. Atkinson's (1976/2008) introduction and notes to Machiavelli's text.

3 See comments by dermatologists critical of Kligman's work, Adewole Adamson and Jules Lipoff, in the *Philadelphia Inquirer*, 2021.

4 The role of phase 1 clinical trials is to determine the safety of new drugs. Because these trials constitute the first time that a new drug is tested on human beings, they are considered the most risky or dangerous of all the trials a new drug must undergo before FDA approval. See Keramet Reiter's (2009) "Experimentation on Prisoners: Persistent Dilemmas in Rights and Regulations." Citing the American Civil Liberties Union, the *Washington Post* (Cohn 1975b) noted that about 10 percent of the nation's two hundred thousand prisoners were involved in prison experiments, but the Pharmaceutical Manufacturer's Association placed that number at 2,400 in 1975 (Steele, 1977).

5 Thomas Kuhn's (1970) seminal work on scientific paradigms defines normal science as "research firmly based upon one or more past scientific achievements . . . that some particular scientific community acknowledges for a time as supplying the foundation for its further practice" (p. 10). These achievements collectively make up the paradigm of a community of scientists. Although in Kuhn's account of normal science, a paradigm is a consciously accepted framework specifying acceptable tools, objects, questions, and directions of research, I am using it more broadly to describe the common occurrence and uncontroversial nature of captive test subject research

6 As Wiegman writes, colonial science first defined race according to geography and climate before shifting its focus to possible biological determinants. In this conceptual move from place-based theories to corporeal evidence, natural history centered skin as the primary indicator of racial difference.

7 I am reminded of Krista Thompson's writings on the visual histories of slavery wherein she cautions against compensatory responses to absences in photographic records. Absence is not lack but constitutes "an intrinsic part of, even representation of, the history of slavery and post-emancipation" (2011, p. 63). The violence of absence extends to representations repeating and redressing it, suggesting that history is "an open site that can be reconfigured both despite and because of the ongoing modes of violence that situate black subjects within modern

regimes of power" (Copeland and Thompson, 2011, p. 6). If, as Thompson argues, we should not allow photographs or archival presence to do the remembering for us, then our analysis generates forms of remembering that are not totally reducible to our objects of study.

8 To name a few here, see Dennis Childs (2015), Michelle Alexander (2010), Colin Dayan (2001), and Loïc Waquant (2002), as well as the groundbreaking work of Angela Davis (1998, 2003).

9 Michael Lynch (1985) writes on the socially and technologically mediated making of scientific artifacts, showing how the everyday work and talk of the laboratory achieve rather than discover the natural scientific order it studies. Avoiding objectivist and anti-objectivist perspectives on science, Karin Knorr-Cetina (1981) similarly argues that science constructs itself and its object, both products of the laboratory's "highly internally structured" practices.

10 Latour and Woolgar describe lab work as a collection of local actions, or "microprocesses." And paradoxically, it is this highly contingent nature of lab work that enables its generalization to the world outside of it.

11 Helen Tilley (2011), for instance, shows how the nineteenth-century "scramble for Africa" positioned the continent as a "living laboratory" for scientific and medical research, social scientific fieldwork, and the development of imperial policy in social engineering and colonial state building. Nancy Leys Stepan's (2001) work on nineteenth- and twentieth-century European representations of the so-called New World charts a visual and geographic grammar of "tropicality" that circumscribed disease, climate, and people within a mythic landscape of exotic, unspoilt, untamed nature. These works, among others, point to a "scientific network" (Latour, 1987) of mobile knowledge objects and inscriptions that disperse the borders of the laboratory proper but in ways that reproduce and rely on the spatial control of non-European bodies. See also works by Warwick Anderson (2003), Duana Fullwiley (2007), Anne Pollock (2012), Troy Duster (2003), Kim TallBear (2013), Evelyn Hammonds (1997), and Sandra Harding (2006, 2011).

12 About another experiment, Brown refused to retell, seemingly too grieved to remember. However, he spent more time describing a series of sunstroke experiments for which he was forced to sit inside a heated pit dug into the ground for extended periods of time.

13 In *The Mismeasure of Man* (1981), Stephen J. Gould has debunked Morton's findings by challenging the rigor and accuracy of his cranial measurements and sampling choices. Gould himself was both a scientist and a historian, and his groundbreaking book on scientific racism exemplifies a uniquely scientific and technological approach to connecting the social context of scientific work with its data and technical practices. For more on Cartwright's work, see Gould, as well as Jonathan Metzl's *The Protest Psychosis* (2009), which details how expressions of black masculinity during the civil rights era (e.g., protest) became increasingly associated with schizophrenia, a clinical diagnosis and popular belief that catalyzed the growth of prisons during a period of deinstitutionalization. Metzl connects

this pathologization of black resistance in the 1960s to Cartwright's work from the antebellum period, which legitimized enslavement by arguing that servility was both the natural state of black people and that it invigorated their naturally poor constitution.

14 Agassiz's daguerreotypes reflect conventions of early scientific illustration then taken up and further developed by eugenicist projects that followed in their wake. Smith situates Agassiz's photographs within a long anthropological tradition of imaging indigenous or colonized subjects and links it to Francis Galton's composite photographs that sought to illustrate biological types. W. E. B. Du Bois would appropriate the representational conventions of both biological racialism and family portraiture to produce a new science of race (or "scientific propaganda"), one that would subvert biological theories by imaging black subjects as visual evidence of shared humanity among the races. See also Jasmine Nichole Cobb's *Picture Freedom: Remaking Black Visuality in the Early Nineteenth Century* (2015), which traces how free blacks deployed popular media including daguerreotypes to found a visual culture of black freedom and that would bear on public understandings of nationhood and citizenship. Nancy Leys Stepan (2001) also analyzes Agassiz's work in colonial tropical science, where he was prominent. Soon after his photographic studies in US plantations, Agassiz traveled to Brazil to continue his work on racial biology. There, Agassiz again commissioned photographs of non-white subjects, this time of natives and mixed-race peoples. His aim was to illustrate how miscegenation could catalyze racial degeneracy.

15 Gould and others (such as Kim TallBear) write extensively on the differences between monogenism and polygenism, two origin theories about human beings. Gould hypothesized a single origin, whereas TallBear argued for multiple lines of descent such that the races constituted separate species. Agassiz and Morton were polygenists, whereas Charles Darwin's famous work supports the monogenist hypothesis. Though Gould disproves Morton's biological determinism to dispute science's claim to objectivity, he emphasizes that such thinking must not be reduced to bad intentions. He writes, "Conscious fraud is probably rare in science. It is also not very interesting, for it tells us little about the nature of scientific activity" (p. 86).

16 As Priscilla Wald (2012) also writes on race and ethics in biotechnology and medicine, "Racist assumptions have historically structured the definitions of the human being on which assertions of inviolable human rights rely" (p. 249).

17 Braun provides a trenchant analysis of race correction in diagnostic tools like the spirometer, which measures lung capacity. During slavery, the spirometer was used to demonstrate the smaller lung capacities of slaves. An assumption now built into the technology through race correction, scientific differentiation of lungs based on race joins other practices where black patients are enjoined to show greater degree of harm to acquire care or compensation for injury.

18 Karla Holloway (2011) compares the phrase *clinical research* with *medical experimentation*, each having very different connotations though they refer to the exact same thing: testing possible treatments on human subjects. Holloway calls this

difference a "deep structure" evoking medicine's history of abusing exploitable populations, thus "return[ing] us to the matter of identities that are institutionally expedient rather than medically relevant" (p. 106). By excavating the deep structure of bioethics, Holloway can bring to the fore the "excesses" or the sociopolitical complexities and contradictions peeled away through the idiosyncratic legal language of bioethics.

19 Brown theorizes the "repeating body" of slavery, both an image and a form of seeing engendered by slavery. Like an afterimage, the repeating body suggests that the culturally rehearsed violence of seeing, representing, and remembering black women is a problem of hypersurveillance, of always being subject to the gaze.

20 For more on medical practice and the institution of slavery, see, for example, Katherine Bankole (1998) and Dea H. Boster (2013).

21 Emphasis added. Statement of principles on the conduct of pharmaceutical research in the prison environment (1975) presented to the National Commission for the Protection of Human Subjects of Biomedical and Behavioral Research. The statement references new FDA requirements created in response to the thalidomide tragedy that occurred between 1957 and 1961. Thalidomide was marketed worldwide to pregnant women as an effective treatment for morning sickness. Thalidomide, however, caused thousands of infants to be born with physical disabilities, prompting the FDA to institute stricter guidelines for medical research—guidelines that would lead to the current US model of drug monitoring, that is, preclinical trials in animals followed by three phases of testing safety and efficacy in humans.

22 Also quoted from the commission's proceedings on January 9, 1975. Acquired through the Bioethics Research Library at Georgetown University.

23 Notable institutions for this kind of research include the Public Health Service Addiction Research Center at Lexington, South Carolina, and the Malaria Research Project at Statesville Penitentiary in Illinois. For forty years, the former studied drug addiction to opiates and alcohol as well as the abuse potential of new pharmacological agents, while the latter experimented on antimalarial treatments and possible vaccines and cures for nearly thirty years. Both closed soon after the formation of the country's first federal taskforce on bioethics, which is discussed further in chapter 3 of this text. See also Welsome (1999), Comfort (2009), and Hatch (2019).

24 Written report by the National Prison Project of the American Civil Liberties Union Foundation, submitted to the National Commission for the Protection of Human Subjects of Biomedical and Behavioral Research, 1976. Acquired through the Bioethics Research Library at Georgetown University.

25 The latter were ascertained in reports by the Prisoners' Rights Council of Philadelphia and the National Commission for the Protection of Human Subjects of Biomedical and Behavioral Research.

26 Coincidentally, the reviewer who instigated the FDA's investigation into Kligman's research at Holmesburg was none other than Dr. Frances O. Kelsey, who

prevented the sale of thalidomide in the US market and catalyzed the making of more stringent clinical practice guidelines that would lead pharmaceuticals to seek out prisoner test subjects (Mintz, 1966, July 23). Calling the FDA ban "absurd" and the object of its grievances a "tiny little matter" (ibid.), Kligman challenged and eventually was able to get overturned the FDA's blocking of his research at Holmesburg.

27 Kligman coined the term *cosmeceutical* in 1984 to classify products resulting from skincare science, though it remains a vague and controversial term in dermatology.

28 In their historical analysis of Agassiz's photographs, Molly Rogers and David Blight (2010) recount that racialist questions about what the human is always accompanied and were later supplanted by those about the beautiful and the good. They write, "Bodies were no longer scrutinized and classified to make sense of the world; now they were judged desirable and detestable for the purpose of maintaining social order" (p. 20).

29 See also Benjamin's edited volume *Captivating Technology: Race, Carceral Technoscience, and Liberatory Imagination in Everyday Life* (2019) examining the intersections of captivity and technology design, showing how the control and surveillance apparatus of the plantation and its aftermaths become the general condition of modern life.

30 Thus, Chun makes room for race as poieses, an agency that can disturb oppressive forms of revealing and creating meaning. Drawing from Donna Haraway's theorization of the cyborg's radical potential, Chun embraces staying with the technology analytic rather than attempting to recuperate race from technology. Similarly, Beth Coleman (2009) sees counternarratives (counter-uses?) of race. As techne, race reveals a power of execution and a mastery of tools that can "be for ill as well as for good; it may become a trap or a trapdoor" (p. 180).

31 Modern science is demonstrable knowledge, and so while it can never be independent of instrumentation, it also shapes the evolution and application of technology in ways that differentiate the tools of science from its culture. This is not to subordinate technology to scientific thinking, and this is certainly not to rehash debates about the contested priority of technology in scientific culture. Don Ihde (1983) introduced this problem by claiming technology's historical and ontological priority in relation to science, which requires the a priori existence of technology to perform its self-conscious inquiry into the laws of nature, and which also relies on technology as a mediating object enabling a physical, sensorial experiencing of the world from which science is then derived. Ihde's thesis was in response to what he believed were idealist theories of technology that construed the latter as both dependent on and constitutive of the application of insights into natural laws. To idealists, as science progressed toward more mechanistic explanations of the world and away from its more speculative roots in classical thought and natural philosophy, it paved the way for technological developments or material constructions like the Industrial Revolution. While I am unconcerned with

this split between technology and science, I recognize that it is no mere semantic quarrel, for what is at stake in this dialogue is the very status of technology as an analytical category and, for scholars, an impetus for field building (e.g., What differentiates a historian of science from a historian of technology?). A permutation of this question on technological priority is manifest in debates on technological determinism and the social construction of technology, and while I am unconcerned, too, with this split, eschewing the nature/culture binary through the race-as-technology analytic still bears on that between technology and culture (not to mention machine/organism, which Donna Haraway's "cyborg" has since complicated). Still, if Ihde's thesis holds, if technology is historically and ontologically prior to science as a culture, then is it also prior to other formations like labor, gender, governance, and economic structure? If not, and one adopts the constructivist approach—that technology and society are co-determining or that technology's impact depends on context—how does this framework then map onto the relationship between technology and scientific culture?

32 This is a reprinted ethnographic account of how individuals are brought into culture through dress and decoration: "The surface of the body . . . becomes the symbolic stage upon which the drama of socialization is enacted" (p. 486). Adorning the body constitutes a kind of drama, because it reveals the tension between communicating the self and conforming the self to cultural expectations. Hence, dress and decoration as expressions of individual tastes and attitudes also materialize the subject's integration into society.

33 In her text, Keeling is using the translation by Charles Lam Markmann.

34 Fanon argued that ontology ignores lived experience (p. 90) but also warned that in taking "into your head to express existence, you will very likely encounter nothing but the nonexistent" (p.116).

35 Fanon writes, "I slip into corners, my long antenna encountering *the various axioms on the surface of things*," (p. 96, emphasis added).

36 Lacking meaning, it can mean anything. Fanon argues that black being is a "toy in the hands of the white man," making direct reference to Sigmund Freud's description of *fort-da*, or "gone-there." Freud had observed his grandson exclaiming these words during play, where the child would throw his toy away ("Gone!"), bring it back to himself ("There!"), only to throw it away again. Freud theorized that this form of early childhood repetition allowed the child to regulate anxiety, here the absence of the mother. In referencing this game, Fanon might suggest the regressive nature of whiteness. But more importantly, he implies an absence at the heart of whiteness that the latter attempts to cover up with its antagonistic relationship to blackness. In this sense, blackness is both absence and its deferral, "at the crossroads between Nothingness and Infinity" (p. 119). And this is another of way of reading the nature/culture binary in Fanon's text, that it is a game of *fort* ("Nature!") and *da* ("Culture!").

37 "If I had to define myself," Fanon says, "I would say I am in expectation; I am investigating my surroundings. . . . I have become a sensor" (p. 99).

38 As Nahum Chandler (2008) writes, "The Negro or the African American as a
problem for thought is both *from* an exorbitance, otherwise than according to the
classical determinations of metaphysics, and *about* an exorbitance, as a problem
within metaphysics as it tries to situate the Negro as within knowledge, according
to philosophy as science" (p. 388, emphasis in the original).

1. THE SKIN APPARATUS

1 In their work, Daston and Gallison (1992) trace the unwavering and unerring
discipline attributed to the camera to the Christian principles of ascetism permeating nineteenth-century Victorian society out of which modern photography
emerged. Of course, photography did not finally settle longstanding controversies
on subjectivity and truth. What it did was technologize and further systematize,
or secularize, ascetism's moralization of self-discipline, raising new questions
within ongoing concerns about the competent interpretation of images and the
accepted criteria for image-making.

   The work of Robert Merton (1938) has elobrated on the ascetic principles
undergirding early European science more broadly, the latter spurred by
seventeenth-century Puritanism, which lent the nascent field of science a legitimacy the latter did not yet have at the time. However, practitioners of science
did not merely incorporate religious values into their work for the sole purpose
of acquiring social acceptance. Rather, they were themselves deeply religious
men who countenanced science as a means of bettering society and of affirming the splendor of God. Merton writes, "The Protestant ethic had pervaded the
realm of science and had left its indelible stamp upon the attitude of scientists
toward their work . . . these worldly activities and scientific achievements manifest the Glory of God and enhance the Good of Man" (p. 115). This integration
of science and religion was manifest in their common privileging of empiri-
cism and utilitarianism. In science, an insistence on practical experiments
reflects Puritan asceticism, that is, an ethic of reason "as a curbing device of
the passions" (p. 120) and of the idleness presumed inherent in contemplation.
Experimentation, in contrast to contemplation, consists of a rational, systematic engagement with and consequently a mastering of the world.

2 Michel Foucault ([1977] 1995) theorizes self-surveillance through Jeremy Ben-
tham's panoptic design for prisons, where the perceived all-seeing eye of authority
figures is internalized by and therefore distributed among the prisoners themselves. See also Foucault's *Power/Knowledge: Selected Interviews and Other Writings* (1980) and *The History of Sexuality, Volume 1* ([1976] 1990) on the differences
between negative (repressive) and positive (productive) power. The latter invests
in the body and thereby produces its own subject (desire and knowledge), coming
to have no rationale or justification outside of this process of investment.

3 Comments made by then chairman of the Department of Dermatology at the
University of Pennsylvania School of Medicine, John R. Stanley, for Kligman's
2010 obituary for the *New York Times*.

4 Delimiting images as objectified scientific practices, the sociological framework on scientific vision described here differs considerably from visual culture analyses theorizing the disembodied, invisible nature of the scientific gaze, which in turn makes its object hyper-present. This hypervisibility is in fact a twin manifestation of invisibility, where reified difference, a form of not-seeing, constitutes the site of knowledge production and subject formation. Where these divergent approaches meet is in their shared critique of science's myth of objectivity, challenging its presumed autonomy to define and interrogate itself and its objects. Neither give science the last word, each in their own way toppling "great man" narratives of scientific accomplishments by revealing the underlying interests and assumptions of research.

5 See Leyden (1991). The dermatology department at University of Pennsylvania has been called "the Mecca for pilgrims who searched for postgraduate scientific training" and Kligman the "magnet" for attracting those students (Plewig, 1991, p. 1415).

6 In 2019, for example, the Society for Investigative Dermatology discontinued using Kligman's name in lectureships and travel awards (Adamson & Lipoff, 2020).

7 The stratum corneum, or horny layer, composes the skin's outermost layer and contains only dead cells. Removing in a stepwise fashion each layer of the corneum using Scotch tape, the researchers found that how quickly and how deeply this skin layer glowed under black light positively correlated with how much dye was applied to the skin and how much time the dye was given to seep into it (Baker & Kligman, 1967).

8 The researchers ascertained that treating the isolated corneum with alkaline buffers enables the visualization of its structural organization without the use of electron microscopy (Christophers & Kligman, 1963).

9 Hypothesizing that a substance in contact with skin caused photoallergic dermatitis, this experiment determined that light only increased the potency of a photosensitive drug that can by itself stimulate an allergic reaction; these kinds of substances are allergens, which are divided into direct contact allergens, or haptens, and photoallergic substances, or photosensitizers (Willis & Kligman, 1968a). Some photosensitizers could remain in the skin for several months and even years after application, catalyzing allergic reactions to the skin under even minimal exposure to light (Willis & Kligman, 1968b).

10 For Kligman's work on foot fungus, see Maibach and Kligman (1962), Rebora, Marples, and Kligman (1973a and 1973b), and Leyden and Kligman (1978). See also Kligman's 1974 comprehensive report to the US Army Medical Research and Development Command, "Sustained Protection Against Superficial Bacterial and Fungal Infection by Topical Treatment."

11 See Robbins (1983).

12 See patents US 4603146 A, US 5998395 A, and US 6228887 B. Kligman had also patented several uses of topical vitamin A and its other derivatives.

13 On the racial and gendered history of skin bleaching, see Shirley Tate's *Skin Bleaching in Black Atlantic Zones: Shade Shifters* (2015). For market research on skin bleaching/lightening products, see the summary of *Futures Market Report*, at https://www.prnewswire.com/news-releases/skin-lightening-products-market -growing-fascination-for-fair-skin-to-spur-the-growth-of-global-market---future -market-insights-665159303.html.

14 Tretinoin creams and gels available today vary in strength, from 0.01 percent to 0.1 percent. Hydroquinone is also currently sold as a topical treatment for melasma, or the appearance of darker skin patches on the face, at preparations of 2 percent and 4 percent. For his experiments, Kligman arrived through trial-and-error at a formula of 0.1 percent tretinoin, 5 percent hydroquinone, and 0.01 percent dexamethasone (a steroid) in an oil-based solution. Kligman's publication refers to prior studies, others' and his own, demonstrating the bleaching effects of hydroquinone and injectable corticosteroids on black skin, acne-prone white skin, and the skin of guinea pigs.

15 Notice that Fanon's discussion on the biological and the intellectual is grounded in sexuality, where black male sexuality is framed not in terms of desirability or even of purity but of violence: rape. This fixation of black men at the "genital level" is what grounds the entire economy of white phobias and fascinations with the sexual in general.

16 Barthes compares the punctum to an injury. It wounds, "like an arrow," but it is itself also a wound located in the image, both the photograph and the viewer speckled with these holes/points.

17 Knorr-Cetina (1981) calls this the "situational contingency" of methods adopted in the lab.

18. I borrow "hanging together" from Annemarie Mol's *The Body Multiple: Ontology in Medical Practice* (2002), which traces how atherosclerosis is enacted in disparate hospital settings—the clinic, the pathology lab, the surgery table, and so on. Mapping the ways atherosclerosis is distinguished and talked about in the hospital setting, Mol addresses how a presumably single object assembles or hangs together through (*not* despite of) the multiple ways it is practiced, only disappearing when its practices, too, disappear. For Mol, this is the ontology of medical practice; reality (of an illness) is situated in everyday conversations, collaborations, and decisions, and thus always existing as "*more than one and less than many*" (p. 82, emphasis hers).

19 In her historical analysis of racial passing in the United States, Samira Kawash (1997) illustrates how the color line came to produce and organize knowledge about identity and the body. Passing disturbs this relationship between accepted visible signs of race and racial identity, making physical traits like skin color an insufficient guarantee of knowledge about the person's identity. We can see this at work in Kligman's preference for using "deeply" pigmented skin, which made the researcher more confident in his knowledge about the lightening effects of his tretinoin formula.

20 Significantly, Goodwin furthers this argument by analyzing how police defendants in the 1991 Rodney King trial were able to secure an acquittal for their beating of Rodney King, which was caught on camera. Submitted in court as evidence, the tape was reframed by the prosecutors not as an instance of brutality and excessive use of force but as an example of the deliberate and lawful execution of standard operating procedure.

21 I borrow "burden of representation" from John Tagg (1993), who charts the making of photographic evidence in the second half of the nineteenth century, when Western powers began seeking greater access to and regulation of social life via new modes of documentary practice. From the modernization of surveillance and police power in France and Great Britain and later to the New Deal's formation of a US welfare state, photography's association with fact and truth was solidified through its role in recording myriad social issues targeted for progressive state intervention.

22 Like Smith, I am referencing W. J. T. Mitchell's explanation (or defense?) of visual culture studies in his article, "*Showing Seeing*: A Critique of Visual Culture" (2002).

23 As discussed in the introduction to this book, histology regularly involves excising tissue samples from a body and freezing or embedding those samples in paraffin wax. Researchers then slice them into thin, translucent sections with knives called microtomes and then place those sections onto glass slides where they are stained before viewing under a microscope.

24 For Cartwright, the scientific gaze is not immune from the pleasures and politics of looking, which accompany this gaze as it becomes dispersed and embodied through specialized instruments of seeing in the lab. Forming privileged sites of constructing knowledge about the human body, these instruments also include the animals made to model human diseases and conditions, composing the "intermediary" or "symbolic" terrain where "human corporeal regulation is carried out" (p. 93).

25 Jared Sexton and Steve Martinot (2003) illustrate how race is often conflated with spectacles of state violence, arguing that efforts to plumb the depths or hidden meanings of antiblack racism overlook the banality of policing and killing black life. They call the spectacular nature of racial violence a "camouflage": "It does not conceal anything; it simply renders it unrecognizable. One looks at it and does not see it" (p. 174). Here, I find this concept of camouflage a helpful approach for comparing photographs and photomicrographs of depigmented skin, and for questioning the recognizability of race in scientific images. For Sexton and Martinot, racism does not contain within it a hidden logic that would dissolve as soon as it is unearthed. Rather, the truth of racism is "that the truth is on the surface, flat and repetitive, just as the law is made by the uniform" (p. 179). I borrow this argument in my own about the visual language of scientific racism.

26 By "fact of blackness," I am referring to a chapter title in the version of *Black Skin, White Masks* translated by Charles Lam Markmann ([1952] 1986). All other

citations in the text refer to the version translated by Richard Philcox, who interprets the original chapter title as "The lived experience of the black man."

27 I am borrowing from Cheryl Harris's (1993) concept of whiteness as property, in the form of material, symbolic, and legal privilege afforded to white people. I find particularly compelling her discussion of whiteness as capacity to "use and enjoy" black bodies in the context of slavery. This instrumentalization of blackness informs my analysis here.

28 I am borrowing "touching feeling" from Eve Kosofsky Sedgwick (2003), who writes on the epistemic possibilities of touch, visuality, and affect. See also the edited volume by Elspeth H. Brown and Thy Phu (2014), which introduces Sedgwick's interventions to the study of photography.

29 See also Jennifer Barker's *The Tactile Eye: Touch and the Cinematic Experience* (2009), which connects touching to other senses involved in spectatorship. Conceptualizing touch at both surfaces and depths, Barker illustrates how spectatorship entails the viewer's "total immersion" in the film viewed.

30 Jonathan Crary (1988) has written on the ways technologies of seeing came to materialize a Western belief in the direct relationship between vision and truth, an epistemological order the camera obscura epitomized during the seventeenth and eighteenth centuries. Promising access to an objective knowledge of the world, the camera obscura's geometrical optics was taken as an infallible, detached vantage point of observation that could displace the unreliable seeing subject. This visual paradigm would change in the nineteenth century, when a greater focus on embodied seeing demonstrated vision's inescapable reliance on anatomy and physiology.

2. SKIN PROBLEMS

1 See Kligman's 2010 obituary in the *Philadelphia Inquirer*.

2 See statistics and calls for change in Perlman, Klein, and Park (2020), Lester and Taylor (2021), and Smith and Oliver (2021). Significantly, these calls mobilize dermatology for fighting racism, referring to summer 2020's mass demonstrations for black life in the aftermath of police shootings, notably of Breonna Tayler and George Floyd. Adamson and Lipoff (2020) call for discontinuing using Kligman's name for professorships, lectureships, and departments.

3 The turn to material agency is multiple, split between new materialism, object-oriented ontology, and speculative realism, with much theoretical overlap between them. Whereas new materialism opposes itself to representationalism, object-oriented ontology and speculative realism interrogate the anthropomorphism in Kantian philosophy and its derivatives in contemporary scholarship. Specifically, they repudiate the correlation made between reality and human thought, contending that objects are not fully contained by human renderings, that they exist apart from us and that, indeed, all there is in the world are objects. Two anthologies evidencing the mounting purchase of this new metaphysics of matter in the field of science and technology studies include *New Materialisms: Ontology, Agency,*

and Politics, edited by Diana Coole and Samantha Frost (2010), and *Material Feminisms*, by Stacy Alaimo and Susan Hekman (2008), both of which include contributions from notable pioneers: Rosi Braidotti, Elizabeth Grosz, Vicky Kirby, Karen Barad, Suson Bordo, among others. Collectively, these approaches view materiality as "excess, force, vitality, relationality, or difference that renders matter active, self-creative, productive, unpredictable" (Coole & Frost, p. 9).

4 Of course, I am referring here to Donna Haraway's (1991) seminal work on the cyborg, which has since grounded numerous feminist writings on the posthuman. For Haraway, the cyborg is an oppositional metaphor against modern binaries of human/non-human and mind/body. The cyborg muddies this division between ideas and things, representing "transgressed boundaries, potent fusions, and dangerous possibilities" (p. 154). It thus decenters the human and reveals the "disturbingly lively" (p. 152) character of matter, obscuring the ontological borders between subject/object enshrined in modern thought. Elaborating on how Haraway's cyborg inaugurated feminism's material, ontological, and posthumanist turn, Cecilia Åsberg (2010) writes that it proposed a "way of unpacking and holding up for inspection some intricate details that really effects all kinds of feminist politics and other democratic issues of science," including "complicated issues of life and healing, death and suffering, of who gets to live, who gets to die and who decides that" (p. 20).

5 Co-presence is a prevalent theme in Bruno Latour's work (1992), which starts from the *relationship* between artifacts and humans rather than theorizing from one category to the other. Latour deems these relationships the "missing masses" of sociological work. So instead of creating "two heaps of refuse, 'society' on one side and 'technology' on the other," Latour focuses on how activities are distributed in a chain of human and nonhuman actors (p. 240).

6 In his comparison of slave systems from ancient to modern societies across the globe, Patterson elaborates on general dishonor and natal alienation as the constitutive elements of social death. General dishonor refers to the totality of the master's brute force, against which the slave held no recourse to self-defense. Natal alienation describes slavery's obliteration of kinship ties, where slaves were considered social nonpersons with a past but no heritage, "genealogical isolates" existing solely as extensions of their masters' power.

7 Kligman would later modify this test by replacing the patch with a round aluminum chamber. Kligman published the results in "The Chamber-Scarification for Irritancy" (1976) and "The Duhring Chamber" (1979).

8 An allergic reaction does not happen during the first exposure to a foreign substance but during the second, wherein immune cells in the skin and lymph nodes have learned to mount a response to the substance after they have become "sensitized" to it or have developed a "memory" of it during its initial presentation.

9 These observations were compared against the results of control patches, which did not contain test agents. The scoring system was later reduced to four grades of allergenicity.

10 Goeffrey C. Bowker and Susan Leigh Star's *Sorting Things Out: Classification and Its Consequences* (2000) examines how categories become tacit, or how the work behind making categories becomes invisible. As ethical choices, categories implicitly and explicitly provide value judgments and become naturalized through the creation of boundary objects, which are made to police and maintain the coherence of communities in between categories without destabilizing the latter.

11 See sebum studies discussed in chapter 1.

12 See Leyden (1991).

13 See Kligman (1991, p. 1375).

14 In the skin lightening experiments, race did not constitute the object of study—the tretinoin formula did. In the maximization test, the immune system did not constitute the object of study—race did. But in both kinds of research, the skin's ability to *visualize* reactions constituted the source of evidence about the objects of study.

15 Allen Hornblum adopted this photograph for the cover of his infamous exposé, *Acres of Skin: Human Experiments at Holmesburg Prison.* In this front cover, much of the photograph is cropped to more narrowly focus on McBride and the test subject. And whereas the original picture meant to convey a positive reading—prisoners volunteer to save lives!—this truncated, zoomed-in image of the medical examination marshals the opposite sentiment, appalling the viewer with visual proof of abuse and exploitation. It pictures the intent behind the book, bringing the prisoner up close to show what lays in the pages beyond it: an "opportunity for those who contributed their bodies to science . . . to tell their side of the story: To explain why they chose to be, and what it is like to be, a Holmesburg guinea pig" (p. 4). Opening his text with the Nuremberg Code and leaving the reader with a closing warning against future abuses, Hornblum positions *Acres of Skin* within a larger discourse of bioethics, which seeks to investigate, regulate, and correct biomedical mistreatment.

Originally made to capture and relay the gains of prison research, the photograph's more recent interpretation now urges the viewer to condemn what it shows. Still, in both cases, it startles and captures attention, which is drawn to the six patches methodically placed onto the test subject's back, orderly and seemingly innocent against the relaxed expanse of his torso. It possesses a striking resemblance to one of the most iconic images of slavery, the late nineteenth-century photograph of runaway slave Gordon. Created for abolitionist consciousness raising, *The Scourged Back* shocks its viewer by giving an unimpeded survey of Gordon's gnarled, twisted flesh, his full back a constellation of knotted shapes and bulbous lines indicating skin and muscle that had been repeatedly torn away by the whip. This image meant to visualize the singular horrors of slavery shares with the photograph of the patch exam a pattern of display: the body turned away from the camera, injury and nakedness seen through the back, the photographed subject unable to look at the harm upon which the viewer gazes. *The Scourged Back*, nonetheless, has been critiqued for its voyeuristic exhibition of pain (Jackson, 2011). Transforming pain into spectacle, *The Scourged Back* renders unrecognizable the banal violence of slavery, mystifying the very same institution it was meant to critique. The new uses of the patch-exam photograph ineluctably resonates with this context of looking, stimulating public outrage (or fascination) by making exemplary the otherwise everyday operations of postwar prison violence, including medical research. When interpreted as testaments of human suffering or determination, the photograph can awaken sympathies or affirmations of atrocity originally denied, but as Susan Sontag (2003) has also cautioned, asking the viewer to sympathize is not the same thing as asking her to think or understand. Tethering the viewer's compassion to their entitlement to look, sympathy and recognition do not necessarily bring one closer to the subject in pain, instead forming a vehicle of sentimentality through which one can fetishize and therefore distance oneself from the pain of others.

16 William Martin's correspondence with the commission, September 3, 1975. Acquired from Georgetown University, Washington, DC.

17 Comments by Kenneth Hardy to the national commission on bioethics, documented in the *Washington Post*, November 16, 1975. Acquired through Bioethics Research Library, Georgetown University.

18 Given the secretive nature of such programs and of imprisonment in general, the absolute number of captive test subjects and the racial breakdown of that population remains unknown. Available reports indicate that white prisoners were more involved in experiments, but, as Hatch argues, the lack of transparency does not make that observation conclusive. Engaging with the same archive, I have also found anecdotal evidence that white prisoners were more often enrolled in "safer" experiments and black prisoners in the more dangerous ones. But as with Hatch, the archive does not provide definitive proof of this. Hatch writes that because black prisoners tended be overrepresented in captive populations, assumptions were made that they then were also overrepresented in testing programs; the archive does not bear this out. My research site, Holmesburg, may be unique in this regard, given that Kligman had confirmed the overrepresentation of black prisoners in his test subject population.

19 I am inverting Weheliye's posthumanist question in his *Habeous Viscus: Racializing Assemblages, Biopolitics, and Black Feminist Theories of the Human*, where he writes, "Why are formations of the oppressed deemed liberatory only if they resist hegemony and/or exhibit the full agency of the oppressed? What deformations of freedom become possible in the absence of resistance and agency?" (p. 2).

20 As Allan Sekula (1986) had also illustrated, the science of imprisonment was thoroughly enabled by new forms of seeing the captive body during the nineteenth century. The panoptic gaze was instrumentalized via photography—indeed, the latter catalyzed developments in the former. This visual culture of criminal science, always already influenced by dominant racial and class attitudes, further instituted both the body and society as objects of bureaucratic control. Deploying

photography to standardize the image of criminality (a "criminal type") and to differentiate it from the body of the law-abiding citizen, early proponents of criminology shared with their contemporary eugenicists a commitment to demographic regulation.

21 See Sarah Ahmed and Jackie Stacey's *Thinking through the Skin* (2001) and Marc Lafrance's (2018) edited special issue "Skin Studies: Past, Present, and Future."

3. THE SKIN OF ARCHITECTURE

1 See www.hauntedhovel.com.

2 Original blog at www.ghosteyes.com. "Holmesburg Prison Essay" at https://www .ipl.org/.

3 A local newspaper describes the conflict between the city of Philadelphia and Holmesburg residents, who wish not only to see Holmesburg prison demolished but also that the city not build another prison in their area, which is home to six others (Waring, 2015).

4 The films are *Condition Red* (1995), *Up Close and Personal* (1996), *Animal Factory* (2000), and *Law Abiding Citizen* (2009). Holmesburg was also the site of locally produced movie, *Against the Night* (2017). The prison site was the filming location of Harrison Smith's *Death House*. Artists who have filmed music videos at Holmesburg include Against the Current.

5 See the statement by curator José Roca (2011).

6 Examples of Holmesburg ruinscapes can be found on the photo-sharing and hosting service Flickr. Here, see portfolios by King Crush, Chandra Lampreich, Vince Herbe, Kurt Tavares, and Peter Woodall.

7 See news stories on Detroit's ruin porn by Mark Binelli (2012) and Mike Rubin (2011) for the *New York Times* and Noreen Malone (2011) for the *New Republic*.

8 In their introduction to the collected essays in *Ruins of Modernity* (2009), Julia Hell and Andreas Schönle argue that ruins constitute a visual archive of buildings and their histories as well as a "*transhistorical* iconography of decay and catastrophe" (p. 1, emphasis mine). For one, ruin gazing is inseparable from imperial explorations and archiving of ancient sites. However, it also generates an abrupt awakening of destruction and wreckage, like that following Hurricane Katrina and 9/11, as well as moral and historical lessons about otherwise senseless destruction. Ruins themselves materialize a "reflexivity of a culture that interrogates its own becoming" (p. 7), a "master trope of modern reflexivity [that] encapsulates vacuity and loss as underlying constituents of the modern identity" (pp. 6–7). The "semantic instability" of ruins hence derives from compensatory symbolic activity meant to fill up this lack of meaning. Ultimately, the meaning of ruins is multiple and does not solely come from a real or imagined past but depends in large part on who looks at them.

9 See Peggy Nelson (2010) at *HiLoBrow*. As Kimberly Juanita Brown (2014) shows, however, putting humans within the frame of disaster does not necessarily negate the violence of representation and the racialized "participatory gaze" (p. 183).

Visual traditions banalizing black suffering make of black bodies "indiscriminate sites of repetitive trauma" (p. 195). Brown writes, "The gaze is an empire, rendered with abandon onto particular bodies, presumably for all time" (p. 186).

10 Dimensions from the 1912 volume of the *Journal of Prison Discipline and Philanthropy* (Pennsylvania Prison Society, 1912).

11 See Kostis Kourelis's (2011) discussion on "Doing Time/Depth of Surface," to be examined in more detail later in the chapter.

12 For more on the history of imprisonment during the Progressive Era, see Norman Johnston's *Forms of Constraint: A History of Prison Architecture* (2000) and *Eastern State Penitentiary: A Crucible of Good Intentions* (1994) cowritten with Kenneth Finkel and Jeffrey Cohen. See also Michel Foucault's *Discipline and Punishment* ([1977] 1995).

Eastern State Penitentiary was decommissioned in 1971 and is now a museum. Its website provides digitized archival materials and sources about the prison. As the museum's audio guide narrated during my visit there, Eastern State's first prisoner was a black man, Charles Williams, and its incarcerated population would remain disproportionately black for the duration of its use, a fact perhaps overshadowed by the prison's history of more notable white captives like Al "Scarface" Capone and "Slick Willie" Sutton. The prison museum now educates its visitors about the building's history and on more current trends in incarceration more broadly, centering the unequal treatment of racial minorities in the criminal justice system. Exhibits and on-site art installations point to this "crisis" of race and incarceration, producing and presenting critique from within the literal structure of their object. The museum makes its audio guide, "The Voices of Eastern State," available online at https://www.easternstate.org/explore/audio-tour.

13 See Aimi Hamraie (2017) on universal design and the ideological underpinnings of the built environment.

14 For more on the relationship between archives and garbage, see Shanks, Platt, and Rathje (2004); and Huyssen (2009).

15 This inverts what Hal Foster (2004) calls the archival impulse in art, where artists use found objects in the archive to comment on the present through the lens of the past.

16 See "Angel of History" from Walter Benjamin's ([1940] 1968) description of Klee's *Angelus Novus*: "His face is turned towards the past. Where *we* see the appearance of a chain of events, *he* sees one single catastrophe, which unceasingly piles rubble on top of rubble and hurls it before his feet. He would like to pause for a moment so fair ... to awaken the dead and to piece together what has been smashed. But a storm is blowing from Paradise, it has caught itself up in his wings and is so strong that the Angel can no longer close them. The storm drives him irresistibly into the future, to which his back is turned, while the rubble-heap before him grows sky-high. That which we call progress, is *this* storm" (pp. 5–6, emphasis his). In language reminiscent of Fanon, Benjamin calls for an explosion

of this cycle of ruin we call history or in which the victor arising from the ruins determines history.

17 Evident in Ginsberg's writing on ruins is a metonymic relation between loss, freedom, and death. In ruins, matter is freed/lost from structural form; form is freed/lost from function, and function is freed/lost from the objective for which the building was originally constructed. What is left is a "skeleton" or "dismembered corpse," a symbol of a "death and dissolution [that] occur to all things throughout endless time" (p. 214). He continues, "Death, our death, is ruin's greatest symbolism" (p. 359). Attempts to recuperate a ruin do not restore it to (a different) life but instead constitute an insistence on one's historicity. Nostalgia, Ginsberg argues, is therefore less about us remembering the past than it is the past remembering us.

18 Barthes terms the photograph a "flat Death," referring at once to the flatness of the image (hence, also his use of "camera lucida"), the flatness (which is to say "ubiquity") of death, and the photograph's capacity to kill and "embalm" its object. Aiming at "lifelike" qualities rather than the life of its object, every photograph occasions "the return of the dead" (p. 16).

19 This observation is part of a larger argument against the ocularcentrism of architectural theory, which has ostensibly foregone an examination of how bodily senses are all implicated in the making, imagining, and experiencing of built environments. See also Joseph Rykwert's (1996) foundational text on representations of the human body in classical columns, and Kent Bloomer and Charles Willard Moore (1977) on the body and body-centered artistry of architecture. On space and place more generally, see Steve Pile (1996) on the relationship between the body in the city and the city (or geography and cartography) in the mind, and Heidi Nast and Pile's (1998) edited volume on the social and spatial relationships between bodies and places. See also Gillian Rose (1993) on the gendering of places and spaces in the geographic imaginary.

20 Loos's championing of unadorned surfaces came from a colonialist stance against the ornamentation of what were considered "primitive" cultures. To Loos and his contemporaries, plain surfaces signaled the progress and civilization of the West, producing "a nexus of metonymic meanings—purity, cleanliness, simplicity, anonymity, masculinity, civilization, technology, intellectual abstractism— that are set off against notions of excessive adornment, inarticulate sensuality, femininity, backwardness" (Cheng, p. 25). Mabel Wilson (1998) fleshes out this same nexus in Le Corbusier's *Radiant City*, whose experimentations with Taylorized urbanism and imaginings of a "gleaming white metropolis" (p. 102) relied on a controlled blackness positioned as a threatening force against social order as well as an alluring, primal spirit that can infuse a "European soul dampened by the chaotic industrialization of the twentieth century" (p. 108). In his designs for *Radiant City*, Le Corbusier thus includes blackness only as performing or laboring figures. Tobin Siebers (2010) also notes the disqualifying aesthetics of Le Corbusier's work and of architectural theory in general, which presumes physical

and mental ability as the normative body around which built environments are designed.

21 See Roca's curatorial statement. Elsewhere, Roca more explicitly connects the monoprints with death: each print "from a distance resembles topography, or a shroud over a dead body" (in Brady, 2012, para. 12).

22 Gómez and González list several art projects involving prisons that they have been influenced by and which use prisons as sites and objects of critical commentary. As previously mentioned, Eastern State Penitentiary also exhibits and curates artworks addressing prison reform and abolition. See list at http://www .philagrafika.org/pdf/Doing-Time-Captured-Bibliography.pdf.

23 I borrow "mural membranes" from Jennie Hirsh's (2011) analysis of the installation, to be discussed more fully later in the chapter.

24 Influential texts on authenticity in the arts include Walter Benjamin's ([1936] 2006) essay "The Work of Art in the Age of Mechanical Reproduction," which addresses the capacity of mass reproduction to depreciate or jeopardize the "uniqueness," "presence," "historical testimony," or "aura" of art work. Art in capitalism is designed for reproducibility, and this loss of authenticity as a criterion for making art displaces the latter from the realm of ritual to that of politics. Benjamin, for instance, warns of the excesses of mechanical reproduction, ascribing the "horrible features of imperialistic warfare" to a "discrepancy between the tremendous *means* of production and their inadequate utilization in the *process* of production" (p. 131, emphasis mine).

Though print constitutes a medium of and for reproduction (see works by Elizabeth Eisenstein [1983] and Michael Warner [1990]), the archival impetus behind Gómez and González's monoprints arguably conserves the aura of the original wall. Departing from the electronic and digital reproduction of ruin photography, the monoprints form a very different kind of ruinscape than those discussed thus far in this chapter.

25 In this way, the monoprints also aestheticize the imbricated histories and regimes of the prison and the museum. Citing Tony Bennett's (1994) work, Fleetwood situates carceral aesthetics in the co-emergence of modern museums and prisons as vehicles and expressions of state disciplinary power at the turn of the nineteenth century: the former for public instruction of proper citizenship, and the other for discarding those refusing this tutelage

26 Christopher Tilley (1994a) provides a brief review of scientific and humanistic approaches to space and place. Quantitative analysis conceptualizes space as a container that can be geographically measured. This container is universal and stands apart from human actions. Others, in contrast, define space as a medium socially produced through human action. Phenomenology, for example, center the body as the mediating point between the human and the world, demonstrating that no space could exist without a body to perceive it. Space provides context for place, which gives a situatedness to the former. By naming space, one gives it a place. Architectural space, Tilley writes, "involves a deliberate attempt to create

and bound space, create an inside, an outside, a way around, a channel for movement. Architecture is the deliberate creation of space made tangible, visible and
sensible" (p. 17).

27 Note the depersonalizing function of "inmate" and the use of prisoner identification numbers in place of names, as well as that of matching prison uniforms.

28 See comments made to Alexis Gilbert (2000, September 20).

29 The problem of meaning behind ruin photography is also apparent in John Szarkowski's critique of *In Prison Air* (2005), a photo series of Holmesburg Prison created by Thomas Roma. Like Christopher's and Gómez and González's ruinscapes,
this photo series shows images of cells in a nearly identical fashion. Recognized
as a "photographer of high talent and conspicuous achievement," Roma nonetheless befuddles Szarkowski as to why he would produce such a photo collection
(para. 5). Szarkowski continues: "This is the same photographer who gave us the
great, free-spirited dogs of Brooklyn, and the great open pastures of Sicily; and
it is not unreasonable to ask why a photographer dedicated (or half-dedicated)
to the cause of freedom should make this extended, serious, hermetic effort to
produce a book of photographs concerning the very essence of subjugation. . . .
Why Roma chose to do this book . . . does not answer the question why the rest of
us . . . should look at Roma's pictures with some attention" (para. 7–9). Whereas
free-spirited dogs and open pastures are assumed to be meaningful images, those
of degenerating prison cells can only prompt confusion. Szarkowski speculates
on possible meanings behind Roma's photographs, wondering if "we should look
at them as a kind of warning [against] our rapidly growing prison sector" (para.
9–10), but ultimately concluding that with such images "we would not (naturally)
really understand what they meant" (para. 19).

30 The words of Gómez and González in full: "We have the greatest respect for the
people who lived there. We had to weigh our emotions, pulling personal and private emotional things. We had to ask ourselves, is this the right thing to do? We
did it with respect, but we had to ask ourselves, is it right? We decided we were
giving voice to prisoners who wanted to be heard" (in Rosenberg, 2012, para. 12).

31 Said Roca of the monoprints, "There is truth to the common adage 'if walls
could talk,' in the sense of being the silent witnesses of what happens over time,
which is physically and metaphorically imprinted in them" (in Rosenberg, 2012,
para. 16).

32 The emergence of European modernism, Wallace argues, involved stealing and
forcing African art-objects into an ethnographic museum imbued with the
ideological and philosophical logic of orientalism, primitivism, imperialism,
and colonialism. And often, the transit of these objects followed directly from
the dispossession of peoples and destruction of entire villages. Yet these objects
also came to influence Western aesthetics (see works by Picasso and Matisse),
demonstrating the "endless recombination of the various elements of [European
and African models of ideal form], to which we owe many of the treasures of
European Modernism found in the museums of the world today" (p. 379). On

the status of African art as either "authentic" or "constructed" category, Wallace similarly notes, "I believe that such concepts as art and Africa are . . . more 'cooked' than raw, and as such many different hands have contributed to the present recipe . . . in which blacks haven't necessarily had more of a notion of what to really make of the presence of African art in European and American museums than most whites" (p. 377). See also collected essays in Tim Barringer and Tom Flynn (1998) discussing the power relations underpinning the collection and exhibition of artifacts coming from colonized peoples.

33 Quoted from 2014 review in www.comicsgrinder.com.

34 Quoted from Tenney's 2014 blog post on the episode. Emphasis added.

35 See also Jenny Sharpe's (2002) writings on the troubled agency and un-narrated everyday lives of slave women, and Michelle Brown's (2013) discussion of prison tourism as another popular form of disengagement with the horrors of penal practice.

36 Miles argues that not all deaths or sites become subjects of ghost stories. Citing the attacks on the Twin Towers, Miles shows how some deaths are considered "off-limits" while others are "fair game."

37 Eastern State Penitentiary, the precursor to Holmesburg Prison, is also converted into a haunted attraction during Halloween to raise funds for the museum's maintenance and collaborative projects with archivists, historians, and artists.

38 Miles borrows ghostwriting from Gayatri Spivak (1995), who reads Marx and Derrida (and Derrida's misreading of Marx) to analyze the disappearance of women from theory and the ways subaltern women in particular enter economies of reproduction—their "peculiar predications of ghostliness" (p. 66). On haunting, see also Avery Gordon (2008), who, like Miles, sees in ghost stories a return of repressed past and present social violations. Though an act in the present, haunting signals the lingering trouble of painful histories, compelling a "something-to-be-done" in the face of continued social violence.

Derrida's (1994) own conceptualization of "haunting" in his critique of Marx defines the ghost as "a hallucination or simulacrum that is virtually more actual than . . . living presence" (p. 32) and a *being-there* (versus a not-there) from which "one cannot distinguish between the future-to-come and the coming-back of a specter" (p. 35). This latter definition is useful for understanding the "ever-present-ness" of the captive that I am describing here, a condition in which history and futurity converge through the figure of the slave. In Derrida's own work, the specter (which is also a play on Marx's "specter of communism") is deployed to analyze discourses situating communism as first a "future" threat and then as a "past" one.

39 Holland's argument here is about muddying disciplinary margins, a "contamination" that allows for a "(re)mastery of the master's tools" (p. 161) as well as the acknowledgement of where those tools and their uses come from. For Holland, interdisciplinarity poses high stakes for feminist critique, whose continued significance to the living conditions of women of color pivot on the impossible

possibility of speaking the center and marginal discourses in the same space (p. 152). This approach resonates with Gordon's earlier calls for interdisciplinary or conceptual and methodological bridgework between the humanities and social sciences. Studying hauntings, as well as establishing haunting itself as a critical practice/praxis, means writing ghost stories, which for a sociologist like Gordon means using the tools and objects of the humanities to do sociological work.

40 Here, Taussig discusses spirits of Cuna theology, spirits whose power can be released only when their material forms—the sacred objects embodying them—are ceremoniously burned. This produces the spirits' "immateriality of appearance," objectified in ash and debris or the "uttermost matter of matter" (p. 135).

41 On his blog, Tenney discloses his PTSD and recurrent anxiety attacks since his near-death encounter.

42 To quote Taussig on the ashes of sacred objects burned, "embodiment as disembodiment" (p. 135).

43 Sara Ahmed (2006) writes on the epistemological value of "failed" orientations, developing a queer phenomenology that reorients perception to objects and people in the background of accepted knowledge.

44 Jenell Johnson (2014) discusses this remembering and forgetting in relation to monstrous figures in medical history like the mid-twentieth-century lobotomist. Also participating in a paranormal tour of one asylum-turned-museum, Johnson argues its pedagogical value on two registers. First, paranormal tours provide historical accounts. And second, because they preserve or return to the monstrous figure in history, they enact "the desire for its punishment, a desire to right what has been wronged" (p. 174).

45 Lindberg has a modest acting career, which here he puts to use.

46 The *Philadelphia Inquirer* reports on hunger strikes occurring in 1938, 1954, 1964, and 1969a.

47 Comments by Rev. Frederick Forrest Powers Jr. to the *Philadelphia Inquirer* on October 21, 1969. Reporting on prison conditions, the *Philadelphia Inquirer* lists Holmesburg's prison population at 1,325 in 1969b and 1,176 in 1971.

48 Abdul R. JanMohammed (2005) formulates a theory of meat in relation to slavery, in which the captive appropriates and therefore wrenches away from the master/ state the latter's monopolization of death and of the ever-present *threat* of death (pp. 10–15).

49 In this phrase, Foucault refers to high rates of recidivism at Attica, a comment Angela Davis (1998) quotes to illustrate the limitations of applying Foucault's theory of discipline and punishment to US carceral formations. Initially centering the white European male subject of the penal system, Foucault himself, as Davis shows, was compelled to rethink his theories to include race. Also interrogating Foucault's assumptions, Joy James (1996) writes how black Americans have historically always been presumed to be incapable of docility, becoming objects of extra-legal mob violence occurring *after* Foucault's assertion on the disappearance of public displays of brutalization.

50 From the 1968 *Davis Report* discussed immediately in next chapter.

51 Jennifer Barker's *The Tactile Eye: Touch and the Cinematic Experience* (2009) relates touching and seeing to eating, tasting, and incorporation.

4. BIOETHICS AND THE SKIN OF WORDS

1 Comments made to the *Philadelphia Inquirer* article by William B. Collins (1968).

2 In addition to these seven states that permitted biomedical research, another five permitted behavioral experiments. From the commission's final report as well as from news stories covering the commission's formation (Cohn, 1976a, 1976b; *Washington Post,* 1978). Newspapers acquired through the Bioethics Research Library at Georgetown University.

3 The DHEW was divided into the Department of Education and the Department of Health and Human Services in 1979.

4 An estimate provided by the pharmaceutical industry and garnered from newspaper accounts of the commission's meetings, which were always open to the public. For example, see Schmeck (1976a, 1976b), Branson (1977), and Steele (1977). News clippings acquired through Bioethics Research Library at George-town University.

5 See *Captive Genders: Trans Embodiment and the Prison Industrial Complex* (edited by Eric A. Stanley and Nat Smith, 2015) and *Prison Masculinities* (edited by Donald F. Sabo, Terry Allen Kupers, and Willie James London, 2001) for closer analyses of how sexual harassment and assault constitute carceral strategies of control relying on and reproducing normative sexuality and gender binaries. See also Danielle L. McGuire's *At the Dark End of the Street, Black Women, Rape, and Resistance: A New History of the Civil Rights Movement from Rosa Parks to the Rise of Black Power* (2010), which illustrate how captivity, from slavery to imprisonment and Jim Crow, constituted apparatuses of sexual terror policing the spatial and ideological separations made between whites and blacks, and in which prisons and law enforcement played a significant role in both enabling and perpetrating sexual violence.

6 As quoted in David Norris's (1987) review of *The Tradition of Excellence: Dermatology at the University of Pennsylvania, 1870–1985.*

7 The 1968 Davis report on sexual assault in the Philadelphia prison system concluded an ambitious two-year humanitarian endeavor to document and redress captive trauma. Throughout, investigators were thwarted by silences, refusals, and general noncooperation by prisoners and prison officials alike. This broad reluctance to speak, coupled with the economic reasons behind acts of "submission," made any clear distinction between sexual coercion and consent near impossible. Moreover, even with the information they were able to gather, investigators struggled with using the analytical exposition required for describing their research. Statistics and the dispassionate jargon of psychiatrists, psychologists, and other social scientists recruited for the study could hardly

encapsulate the horrors spelled out by captive testimony. Ultimately, investigators eschewed their professional vocabularies to publish verbatim the deeply painful accounts given by survivors, writing, "In an early draft of our report, an attempt was made to couch this illustrative material in sociological, medical, and legal terminology less offensive than the raw, ugly language used by the witnesses and victims. This approach was abandoned. The incidents are raw and ugly. Any attempts to prettify them would be hypocrisy" (p. 9). The report did not prettify its findings, but its conclusions did mine the raw and ugly narratives for directives on improving prison operations, including the immediate removal of prison guards acting as middle-men in what one local newspaper called "sex corruption" (Collins, 1968). Echoing this sensationalized language in its description of Holmesburg's research program, the report called payment received from research participation "the equivalent of a millionaire's income," which exacerbated power hierarchies among prisoners.

8 For more on the long history of bioethics, see *The Birth of Bioethics* (1998) by Albert R. Jonsen, who had been a member of the commission and thereafter a leading figure in bioethics. As a regulatory term, *bioethics* was itself first articulated in the 1960s, evolving from principles of conduct in Western medicine dating back to the Hippocratic Oath's mandate of doing no harm. In a broad attempt to professionalize medicine or to bring cohesion and respectability to the field, the American Medical Association adopted and modernized this classical text for its code of conduct in 1847, revising it four more times, the last in 1966, before the commission would be established. For a longer treatise on medical ethics, see also Jonsen's *A Short History of Medical Ethics* (2000), a comprehensive exploration of ethics in Western medicine from the classical ages to modern American science.

9 A major concern brought up during the commission's meetings was the feasibility of proposed timelines for each issue they were to take on.

10 With the exception of Dorothy I. Height, who was president of the National Council of Negro Women and an activist in women's rights and civil rights, all other members of the commission worked in academia, medicine, or government: Kenneth J. Ryan (chairman), Joseph V. Brady, Karen Lebacqz, Robert E. Cooke, David W. Louisell, Donald W. Seldin, Albert R. Jonsen, Elliot Stellar, Patricia King, and Robert H. Turtle.

11 Another example of the hyper-public notice Holloway describes is the connotative difference between "clinical research" vis-à-vis "medical experimentation," each phrase having very different meanings, though denotatively they mean the exact same thing: testing possible disease treatments on human subjects. Unlike clinical research, medical experimentation evokes medicine's history of abusing exploitable populations, thus "return[ing] us to the matter of identities that are *institutionally expedient* rather than medically relevant" (p. 106, emphasis added). Extending this approach to "vulnerability," Holloway shows, reveals its embeddedness in social inequities rather than in medical markers; vulnerability is not located in the body but enacted through visible economies of the social.

Departing from bioethics' focus on legal and medical language, Holloway sees in fictional narratives a means for exposing and engaging with the deep structures of bioethical terms, a way of restoring the sociopolitical complexities and contradictions lost in the solution- and accuracy-oriented paradigm of bioethics.

12 From the commission's meeting minutes on July 26, 1975, acquired through the Bioethics Research Library at Georgetown University. The commission addressed prison experimental programs following their deliberations on living fetuses, a subject group that the commission was specifically ordered to address first (Jonsen, 1998). Examining the cultural and political implications of this prioritization of the fetus—the fetus as the inaugural subject of bioethics—and its relationship to other vulnerable subjects is beyond the scope of this chapter. For critiques on fetal personhood, see Rosalind P. Petchesky's "The power of visual culture in the politics of reproduction" (1987), Monica Casper's *The Making of the Unborn Patient* (1998), and Val Hartouni's *Cultural Conceptions* (1997). For critiques on the (white) "child" and futurity, see Lee Edelman's *No Future* (2004). For a critique of Edelman, see José Esteban Muñoz's *Cruising Utopia* (2009). Others have also written on the intersections of captivity and disability, which is assumed in this chapter rather than explicitly providing a theoretical roadmap for this connection. For such a roadmap, see "Abled-Bodied Slave" (2019) in the *Journal of Literary and Cultural Disability Studies*. See also Liat Ben-Moshe, Chris Chapman, and Allison Carey's edited volume *Disability Incarcerated* (2014), the essays of which broaden the meaning of "incarceration" to encompass spatial control over disabled bodies, and which address the disabling conditions of imprisonment and the prison's expanding role in mental health services.

13 See Carlson, Boyd, and Webb (2004) for a longer discussion on the evolution of the Declaration of Helsinki.

14 "Report and Recommendations: Research Involving Prisoners," 1976, Department of Health, Education, and Welfare, p. 5. In a letter to the city's governor, the commission had also expressed their alarm toward Jackson Prison's "sheer inadequacy" and "deplorable condition" of medical services provided to prisoners. Correspondence dated December 4, 1975, and acquired through the Bioethics Research Library at Georgetown University.

15 Several speakers at this conference, among them Dr. L. Alex Swan of Fisk University, refused to even propose or consider ways of instituting prisoners' informed consent, asserting instead that the whole prison system be dismantled.

16 Written report by the National Prison Project of the American Civil Liberties Union Foundation, submitted to the National Commission for the Protection of Human Subjects of Biomedical and Behavioral Research, 1976. Acquired through the Bioethics Research Library at Georgetown University.

17 National Commission for the Protection of Human Subjects of Biomedical and Behavior Research, 1976, p. 35.

18 Comments by Michael Filip, Jack London, and Vern Wixom to the *Washington Post* and the *Detroit News*, 1975. Articles acquired through Georgetown University.

19 These recommendations were not codified in full by DHEW, which considered too vaguely defined the commission's language on "compelling" reasons for using prisoners in addition to its description of the scientific "need" of research. For DHEW, such unclear definitions about what counted as "compelling" or "need" did not allow for making clearly-defined regulations, exemplifying the tension between the technical language of the law and the moral statements character-istic of bioethical discourse. Buttressed by public outcry, the commission's final report nonetheless helped terminate remaining therapeutic and non-therapeutic research conducted in prisons, a move already spreading at the state level. From DHEW spokesman, printed in Steele, 1977. Acquired through the Bioethics Research Library at Georgetown University.

20 Comments by commission chairman Kenneth J. Ryan to the *Washington Post* (Cohn, 1976a). Acquired through the Bioethics Research Library at Georgetown University.

21 See Branson (1977).

22 The commission's statement was publicized in Cohn (1976b).

23 Comments published Carson (1975).

24 Setting the terms of contestation, both its content and horizons, assumptive logic revitalizes itself through the conflictual narratives it founds; here, competing statements from the commission and the bureau on the role of research in penal facilities. Wilderson also calls this logic a "parasitic relationship" between the captive body and its representations. To Wilderson, this relationship is thoroughly anti-dialectic, displacing Hegel's theory of the death struggle between master and slave. Wilderson counters relational ontologies (also underlining more contem-porary theories about material life) with the non-relational or non-ontological status of the captive, that is, the latter's position as an enabling vehicle for relations between masters or between freemen. He writes, "This violence which turns a body into flesh, ripped apart literally and imaginatively, destroys the possibility of ontology because it positions the black in an infinite and indeterminately horrify-ing and open vulnerability, an object made available (which is to say fungible) for any subject" (p. 38).

25 Examples include the ways hospital patients are often encouraged to talk through their physical suffering (like the treatment of serious burns), though Anzieu did not argue that this speech works by diverting the patient's atten-tion away from sensations of pain. Instead, the skin of words symbolically reassembles some of the defensive mechanisms of the real skin, which, having been damaged or penetrated, cannot proffer feelings of psychic and physical wholeness. For Anzieu, bodily pain is this loss of corporeal continuity that the skin of words cannot tangibly replace but can subjectively make more bearable. Anzieu theorized the skin-like activity of words using Emmanuelle Moutin's clinical observations of patients with third-degree burns, where the pain of burn treatment is described as "a state of shock," an excruciating process whereby dead and dying skin is awash in caustic fluid and then summarily rent

from the body piece by piece. The observations did not address the patient's psychic state directly, suggesting that the physical occurrence of shock suf-ficiently reflects or stands in for their subjective experience: both the body and the ego disintegrating at the site of the skin. Anzieu, moreover, compares this anguished patient to the helplessness of a newborn baby, "exposed to the ag-gression of the outside world" (p. 202), describing the all-consuming torment of burn treatment as a process of re-infantilization, or a momentary regression in psychic development that brings the ego back to its earlier indistinguishable, formless state.

26 Though critical of Jacque Lacan's emphasis on the linguistic structure of the unconscious, perceiving it as leaving the body by the wayside, Anzieu's skin of words retains Lacan's early writings on the split subject, a subject divided between one that is fractured but looking and another fully contained but only seen. Put another way, Anzieu's skin of words introduces a tactile dimension to Lacan's mirror stage, where a subject lacking in bodily mastery (such as an infant) comes to imagine itself whole or coherent when, upon gazing into a mirror, sees itself as unified body. Anzieu purposely departed from this kind of visuality theorized in Lacan's writings, whose theoretical leanings towards semiotics and linguistics, Anzieu believed, overlooked the corporeal foundations of the psyche. Thus, for Anzieu, the skin of words is primarily a seamless membrane for touching and feeling rather than for seeing and imaging. But despite Anzieu's digression from Lacan, the skin of words preserves or reproduces this mirror function, imbuing the gaze with tactile powers. The eye is contiguous with the skin, where seeing is also a grasping, stroking, or molding. For Anzieu, not only is the skin of words lo-cated in the mind as a kind of shield against psychic intrusion, but it also becomes embedded in and supported by the body's physical perimeters as an organic, palpable structure of defense.

27 Diana Medlicott (1999) writes that the high rates of depression and suicide in prisons follow from the prisoner's loss of spatial and temporal autonomy. Time in particular becomes a source of suffering for prisoners, whose social lives are effectively suspended even as their biological aging progresses (a slow death).

28 July 26, 1975, commission meeting minutes, acquired through the Bioethics Re-search Library at Georgetown University.

29 Now Michigan State, Jackson State was then the nation's largest prison and home to one of more extensive programs of prison-based medical experimentation. At the time of the commission's visit, Jackson State held 5,200 prisoners, eight hundred of whom were serving as test subjects. Over the previous decade, it had been the location of numerous drug trials run by the pharmaceutical manufactur-ers, Parke-Davis (now Pfizer) and Upjohn (whose brands were later split among Pfizer, Monsanto, and Johnson & Johnson).

30 First name only. The prisoner's last name and number is withheld to protect his identity. Correspondence between John and the commission acquired through the Bioethics Research Library at Georgetown University.

31 Nikolas Rose (2007) theorizes how "life itself" becomes both the target and local instrument of power and formulates a "somatic ethics" that could intervene in these more insidious and diffuse processes of domination. Significantly, Rose includes in this somatic ethics a critique of contemporary research in prisons. The prison bureau and the commission did agree on one kind of scientific research to be allowed in penal facilities, which is still performed today: the prisoner as the focus of study. Biological clues of criminality are increasingly garnered through genetic and neurological tools mobilized for determining guilt, responsibility, and sentencing in courtroom trials. Whole new areas of expertise in biocriminality are generating innovative ways of identifying, examining, and controlling these "anti-citizens," all to protect society, as well as to protect prisoners from themselves.

32 From a statement to the *New York Times* (1975) by DHEW Deputy Assistant Secretary Dr. James Dickson III.

33 Comments published in Carson (1975).

34 Epstein mentions the Tuskegee study in his writings on medical research to historicize what he calls the inclusion-and-difference paradigm, a set of modern ideologies, practices, and institutions that claim the oppressive standardization of the white male subject in research and the urgency of including other social groups that have been purportedly understudied or largely excluded from medical tests: women and non-whites. Pointing to health disparities and to problems of extrapolating data from homogenous test subject pools, proponents of this "biomulticulturalism" use the social group as biomedicine's unit of analysis, which displaces (and sits between) the individual person and the abstract, universal subject of medicine. Contrary to this position, Epstein shows that the history of Western medical science has in fact shown a preoccupation with difference—sex, gender, race, ethnicity—especially in dominant medical theories about group superiority and inferiority, theories that sought biological explanations for social inequalities. In the United States in the eighteenth and nineteenth century, for example, the standard test subjects were black people, who, because of slavery and the failures of Reconstruction, were particularly vulnerable to medical science. Although blacks were considered biologically inferior to whites, and although this view undermined attempts to then generalize data gathered from research on black test subjects, the sheer "availability" of blacks trumped most other practical considerations. The "standard human" in medical science is, hence, subject to historical and technological changes, such as the introduction of statistics and quantitative analyses in medicine during the twentieth century that produced the white male as a normative medical category.

35 From a summary meeting transcript about Jackson State Prison. Acquired through the Bioethics Research Library at Georgetown University. Meeting held on November 15, 1975. An article by philosopher Samuel Gorovitz (1976) in the *Journal of Pharmaceutical Sciences* is suggestive of this conceptual move, encouraging the commission to assess the dependency of captive populations against "the broader backdrop of universal human dependency," because at bottom "all

persons are dependent and constrained in various ways" (p. 4). For Gorovitz, coercion is a general and inevitable human condition, differentiated across individual experiences but common to all nonetheless.

36 The psyche, Anzieu argued, coheres through undamaged biological skin acting as its mediating and containing support, and where impairments to this support, such as in the case of burn victims, can lead to a weakened and vulnerable ego. He theorized the skin of words at the individual level, and it is here that its psychic power is most apparent. In the burn unit, talk between patient and caregiver performs palliative work: "Suddenly [the patient] cried out to me, almost aggressively, 'Can't you see I'm in pain! Say something, anything, but talk . . . talk!' I already knew from experience the connection between a bath of words and the soothing of pain. Signaling discreetly to the nurse to keep still, I set about getting the girl to *talk about herself*, steering the conversation towards comforting subjects: her family, her environment, in short, her sources of emotional support" (p. 204, emphasis added). Summoning pleasurable thoughts, those experiences prior to the patient's injury, the skin of words draws its regenerative powers, or its capacity to symbolically substitute for missing skin, from the affective dimensions of touch. The skin of words quiets the shock of burn treatment by replacing physically damaged skin, the ego's original support structure, with an integumentary system affording the patient "symbolic equivalents of the gentleness, suppleness and appropriateness of contact [they have] to give up when touching becomes impossible, prohibited or painful" (p. 204). Pleasant thoughts produce words that soothe by representing the patient to herself in her secure, unharmed state. This observation is instructive on two accounts. First, protection from harm in the physical sense, the reestablishment of physical boundaries, entails interventions at the level of objective reality. Second, implicit in Anzieu's preoccupation with restoring the ego, in which injury is formulated only through its negative relation to the self, is the central role difference plays. The presence of the other, a caregiver invested in shoring up this difference, remains imperative to helping the ego regain its skin or its difference from the other, as seen when the burn patient calls out to or speaks with the caregiver, supplying them with the personal stories essential to rebuilding and re-presenting the self apart from its surroundings. The skin of words in Anzieu's work indexes an ontological break indispensable to forming and representing difference, its restorative potential laying in its capacity to reinscribe or recover *difference* as felt experience and visual relationship between self and other.

CODA

1 "Ex-Inmates Sue Penn and Kligman over Research," *Pennsylvania Gazette*, February 1, 2001, https://www.upenn.edu/gazette/.

2 "William Bradford vs City of Philadephia et al.," 2007, Civil Action No. 06-CV5121, United States District Court for the Eastern District of Pennyslvania.

3 See Rossi and Rossi (2003).

4 As cited in Harkness (1996). This is testimony of German physician and medical historian Dr. Werner Leibbrant, who was persecuted by the Nazis and who denounced both US and Nazi experiments on captive populations. Here, the defense was appropriating his beliefs to recast Nazi medical abuse as standard practice.

5 Discussed in a 1975 article from the *Washington Post* and acquired from the Bioethics Library at Georgetown University.

6 From federal records obtained by the *Philadelphia Inquirer* in 1979, the Army's $386,486 contract with Penn—valued at over $3 million today—had been the largest dollar value ever paid for human drug testing.

7 See article by Aaron Epstein (1979) written for the *Inquirer*, which acquired relevant Army documents via the Freedom of Information Act. These documents, the article states, included reports from Edgewood Arsenal, Maryland, whose scientists were seeking collaborative endeavors with nonmilitary, ideally university-affiliated research programs close to their base. Names of prisoners involved and of compounds used were redacted, the latter simply referred to as numbered agents such as "AGENT 1-H" or "AGENT CAR 302,196." Finally, the documents did "not reveal what the Army did with the Holmesburg test results."

8 This returns us to Wiegman's definition of skin as the visible epistemology of race (discussed in the introduction). As a political category, race's connection to the skin has nothing to do with the biology of skin color but how skin color assumes social and political meanings.

9 Quoted from a brief journal commentary celebrating Kligman's seventieth birthday.

10 This loss of physical boundaries grounds the ungendering of captive flesh, against which gender (the body) could be understood.

11 The vestibule makes an appearance across several of Spillers's writings in addition to the one cited here, "Mama's Baby, Papa's Maybe: An American Grammar Book." These essays include "Interstices: A Small Drama of Words," "'All the Things You Could Be by Now, If Sigmund Freud's Wife Was Your Mother': Psychoanalysis and Race," "'The Permanent Obliquity of an In(pha)llibly Straight': In the Time of Daughters and Fathers," and "Moving on Down the Line: Variations on the African-American Sermon." In each case, the vestibule is made to suggest the symbolic and material efficacy of incapacity.

12 Quoted from *Newsday*, August 28, 1977.

13 The public resource Clinicaltrials.gov provides information and statistics regarding publicly and privately funded clinical studies.

14 As Adriana Petryna's (2007) writings on clinical trials show, there is an element of exoticism present in contemporary patient recruitment, wherein so-called treatment-naïve populations are more desirable than "treatment saturated" ones in the United States, because the latter's pharmaceuticalized bodies generate drug-drug interactions that might nullify or influence drug test results. Inevitably, this sort of biological purity untainted by pharmaceutical drug-use is more often a result of the inaccessibility of treatments.

15 Phrase from www.assatashakur.org.

16 Serres's thanatocracy resonates with the "body-machine-image complex" that, Mark Seltzer (1997) shows, collapses technologies of atrocity with representations of pain in the making of twentieth-century wound culture. In his sociohistorical study of modern hygiene in late nineteenth to early twentieth-century France, Bruno Latour (1988) deploys metaphors of war and peace to translate Louis Pasteur's scientific successes into relations of weak and strong forces. For Latour, discussing science through the language of war enables one to see the enterprise as precisely a network of *relations of force*. Like war, science celebrates great men and jettisons many other actors—human and nonhuman—that help realize victory—here, a victory over microbes. But like war, science is also strategic, involving uncertainty, drift, confusion, and compromise. Achieving victory over the enemy—be it nations or bacteria—is never buttressed solely by rational planning but also implicates violence, politics, and polemics. Using the language of war to follow simultaneously the context and the technical content of science, Latour maps out a network of associations between multiple actors and forces that move the science of hygiene. For more on authoritarian and democratic technics, or how politics are designed or embedded into technologies, see Langdon Winner (1986), Lewis Mumford (1964, 1970), Peter Kropotkin ([1974] 1998), and Murray Bookchin (1982).

17 I am also reminded of Ruha Benjamin's (2016) politics of refusal, wherein a test subject's choice not to participate in an experiment constitutes the starting material for recreating bioethics.

18 See Montgomery (1971).

BIBLIOGRAPHY

Aarons, Dick. 1964. "Fed-up Inmates Won't Eat, Battle Guards." *Philadelphia Inquirer*, December. *Philadelphia Inquirer* collections, Special Collections Research Center (K 7–382 7–383), Temple University, Philadelphia, PA.

Abel, Elizabeth. 2014. "Skin, Flesh, and the Affective Wrinkles of Civil Rights Photography." In E. Brown and T. Phu, eds., *Feeling Photography* (pp. 93–125). Durham, NC: Duke University Press.

Adamson, Adewole, and Jules Lipoff. 2021. "Penn Must Cut Ties with Dr. Albert Kligman, Who Conducted Unethical Human Research on Black Men." *Philadelphia Inquirer*, January 19. https://www.inquirer.com/

Ahmed, Sara. 2006. *Queer Phenomenology: Orientations, Objects, Others*. Durham, NC: Duke University Press.

Ahmed, Sara, and Jackie Stacey, eds. 2001. *Thinking through the Skin*. New York: Routledge.

Alaimo, Stacy, and Susan J. Hekman, eds. 2008. *Material Feminisms*. Bloomington: Indiana University Press.

Alexander, Michelle. 2010. *The New Jim Crow: Mass Incarceration in the Age of Colorblindness*. New York: The New Press.

Anderson, Warwick. 2003. *The Cultivation of Whiteness: Science, Health, and Racial Destiny in Australia*. Durham, NC: Duke University Press.

Antosh, Lou. 1974. "Medical Testing Lab Closed at Holmesburg." *Philadelphia Bulletin*, January 29. The Bulletin Collection, Special Collections Research Center (Box100), Temple University, Philadelphia, PA.

Anzieu, Didier. 1989. *The Skin Ego*. C. Turner, trans. New Haven, CT: Yale University Press.

Baker, Courtney R. 2015. *Humane Insight: Looking at Images of African American Suffering and Death*. Champaign: University of Illinois Press.

Baker, Harvey, and Albert M. Kligman. 1967. "A Simple In Vitro Method for Studying the Permeability of the Human Stratum Corneum." *Journal of Investigative Dermatology*, 48(3), 273–274.

Bankole, Katherine. 1998. *Slavery and Medicine: Enslavement and Medical Practices in Antebellum Louisiana*. New York: Taylor & Francis.

Barad, Karen. 2007. *Meeting the Universe Halfway: Quantum Physics and the Entanglement of Matter and Meaning*. Durham, NC: Duke University Press.

Barker, Jennifer M. 2009. *The Tactile Eye: Touch and the Cinematic Experience*. Berkeley: University of California Press.

Barringer, Tim, and Tom Flynn, eds. 1998. *Colonialism and the Object: Empire, Material Culture and the Museum* (Vol. 2). London: Routledge.

Barthes, Roland. 1981. *Camera Lucida: Reflections on Photography*. Richard Howard, trans. New York: Hill and Wang.

Ben-Mosche, Liat, Chris Chapman, and Allison C. Carey, eds. 2014. *Disability Incarcerated: Imprisonment and Disability in the United States and Canada*. New York: Palgrave McMillan.

Benjamin, Ruha. 2009. *Race after Technology*. Medford, MA: Polity Press.

———. 2016. "Informed Refusal: Toward a Justice-Based Bioethics." *Science, Technology, and Human Values*, 4(6), 967–990.

———, ed. 2019. *Captivating Technology: Race, Carceral Technoscience, and Liberatory Imagination in Everyday Life*. Durham, NC: Duke University Press.

Benjamin, Walter. 1968. "Theses on the Philosophy of History." In Hannah Arendt, ed., *Illuminations* (pp. 253–264). New York: Schocken. (Original work published 1940)

———. 2006. "The Work of Art in the Age of Mechanical Reproduction." In J. Morra and M. Smith, eds., *Visual Culture: Critical Concepts in Media and Cultural Studies* (pp. 114–137). London: Routledge.

Bennett, Jane. 2010. *Vibrant Matter*. Durham, NC: Duke University Press.

Bennett, Tony. 1994. "The Exhibitionary Complex." *Culture/Power/History: A Reader in Contemporary Social Theory*, 127(2004), 521–547.

Binelli, Mark. 2012. "How Detroit Became the World Capital of Staring at Abandoned Old Buildings." *New York Times*, November 9. http://www.nytimes.com/

Bloomer, Kent C., and Charles W. Moore. 1977. *Body, Memory, and Architecture*. New Haven, CT: Yale University Press.

Bookchin, Murray. 1982. *The Ecology of Freedom: The Emergence and Dissolution of Hierarchy*. Palo Alto, CA: Cheshire Books.

Boster, Dea H. 2013. *African American Slavery and Disability: Bodies, Property, and Power in the Antebellum South, 1800–1860*. New York: Routledge.

Brady, Shaun. 2012. "Two Spanish Artists Remove and Preserve Inmates' Graffiti from Holmesburg Prison." *Philly.com*, January 25. http://articles.philly.com/

Branson, Roy. 1977. "Why Use Prisoners for Drug Testing?" *Washington Post*, July 14. The National Commission for the Protection of Human Subjects of Biomedical and Behavioral Research, 1974–1978, Bioethics Research Library (Box 17, Folder 12), Georgetown University, Washington, DC.

Braun, Lundy. 2014. *Breathing Race into the Machine: The Surprising Career of the Spirometer from Plantation to Genetics*. Minneapolis: University of Minnesota Press.

Brown, Elspeth H., and Thy Phu. 2014. *Feeling Photography*. Durham, NC: Duke University Press.

Brown, John. 1885. *Slave Life in Georgia: A Narrative of the Life, Sufferings, and Escape of John Brown, A Fugitive Slave, Now in England*. L. A. Chamerovzow, ed. London: W. M. Watts, Crown Court, Temple Bar.

Brown, Kimberly Juanita. 2014. "Regarding the Pain of the Other: Photography, Famine, and the Transference of Affect." In Elspeth Brown and Thy Phu, eds., *Feeling Photography* (pp. 181–203). Durham, NC: Duke University Press.

——. 2015. *The Repeating Body: Slavery's Visual Resonance in the Contemporary.* Durham, NC: Duke University Press.

Brown, Michelle. 2013. "Penal Spectatorship and the Culture of Punishment." In David Scott, ed., *Why Prison?* (pp. 89–107). Cambridge, UK: Cambridge University Press.

Browne, Simone. 2015. *Dark Matters: On the Surveillance of Blackness.* Durham, NC: Duke University Press.

Campt, Tina M. 2017. *Listening to Images.* Durham, NC: Duke University Press.

Carlson, Robert V., Kenneth M. Boyd, and David J. Webb. 2004. "The Revision of the Declaration of Helsinki: Past, Present and Future." *British Journal of Clinical Pharmacology,* 57(6), 695–713.

Carson, Norman. 1975. "Prison Research Testimony Scored." *US Medicine,* 1(10), October 15. The National Commission for the Protection of Human Subjects of Biomedical and Behavioral Research, 1974–1978, Bioethics Research Library (Meeting 12, Box 3), Georgetown University, Washington, DC.

Carson, Paul E. 1975. Letter to Kenneth J. Ryan, August 7. The National Commission for the Protection of Human Subjects of Biomedical and Behavioral Research, 1974–1978, Bioethics Research Library (Meeting 11, Box 3), Georgetown University, Washington, DC.

Cartwright, Lisa. 1995. *Screening the Body: Tracing Medicine's Visual Culture.* Minneapolis: University of Minnesota Press.

——. 2014a. "Topographies of Feeling." In Elspeth Brown and Thy Phu, eds., *Feeling Photography* (pp. 93–125). Durham, NC: Duke University Press.

——. 2014b. "Visual Science Studies: Always Already Materialist." In Annamaria Carusi, Aud Sissel Hoel, Timoth Webmoor, and Steve Woolgar, eds., *Visualization in the Age of Computerization* (pp. 297–324). New York: Routledge.

Carusi, Aud Sissel Hoel, Timoth Webmoor, and Steve Woolgar, eds. 2014. *Visualization in the Age of Computerization.* New York: Routledge.

Casper, Monica J. 1998. *The Making of the Unborn Patient: A Social Anatomy of Fetal Surgery.* New Brunswick, NJ: Rutgers University Press.

Chandler, Nahum D. 2008. "Of Exorbitance: The Problem of the Negro as a Problem for Thought." *Criticism,* 50(3), 345–410.

Cheng, Anne Anlin. 2011. *Second Skin: Josephine Baker and the Modern Surface.* Oxford: Oxford University Press.

Childs, Dennis. 2015. *Slaves of the State: Black Incarceration from the Chain Gang to the Penitentiary.* Minneapolis: University of Minnesota Press.

Christopher, Matthew. 2014. *Abandoned America: The Age of Consequences.* Versailles: Jonglez Publishing.

Christophers, Enno, and Albert M Kligman. 1963. "Preparation of Isolated Sheets of Human Stratum Corneum." *JAMA Dermatology,* 88(6), 702–705.

Chun, Wendy. 2009. "Introduction: Race and/as Technology; or, How to Do Things to Race." *Camera Obscura, 70*(24). https://doi.org/10.1215/02705346-2008-013

Civil Rights Congress. 1970. *We Charge Genocide: The Crime of Government against the Negro People.* New York: International Publishers. (Original work published 1951)

Cobb, Jasmine N. 2015. *Picture Freedom: Remaking Black Visuality in the Early Nineteenth Century.* New York: New York University Press.

Cohn, Victor. 1975a. "Inmates Oppose Experimental Halt, Assert Right to Volunteer." *Washington Post,* November 16, 1975. The National Commission for the Protection of Human Subjects of Biomedical and Behavioral Research, 1974–1978, Bioethics Research Library (Box 4, Folder 5), Georgetown University, Washington, DC.

———. 1975b. "Medical Research on Prisoners, Poor." *Washington Post,* February 20. The National Commission for the Protection of Human Subjects of Biomedical and Behavioral Research, 1974–1978, Bioethics Research Library (Meeting 16, Box 5), Georgetown University, Washington, DC.

———. 1976a. "Inmate Research Code Urged." *Washington Post,* June 14. The National Commission for the Protection of Human Subjects of Biomedical and Behavioral Research, 1974–1978, Bioethics Research Library (Box 9, Folder 6), Georgetown University, Washington, DC.

———. 1976b. "Prisoner Test Ban Opposed." *Washington Post,* March 14. The National Commission for the Protection of Human Subjects of Biomedical and Behavioral Research, 1974–1978, Bioethics Research Library (Box 7, Folder 22), Georgetown University, Washington, DC.

———. 1976c. "U.S. Bans Drug Tests on Convicts." *Washington Post,* March 2. The National Commission for the Protection of Human Subjects of Biomedical and Behavioral Research, 1974–1978, Bioethics Research Library (Box 7, Folder 22), Georgetown University, Washington, DC.

Coleman, Beth. 2009. "Race as Technology." *Camera Obscura, 70*(24), 177–207.

Collins, Helen C. 1976. Letter to the National Commission, November 22. The National Commission for the Protection of Human Subjects of Biomedical and Behavioral Research, 1974–1978, Bioethics Research Library (Box 11, Folder 7), Georgetown University, Washington, DC.

Collins, John J. 1976. Letter to the National Commission, November 22. The National Commission for the Protection of Human Subjects of Biomedical and Behavioral Research, 1974–1978, Bioethics Research Library (Box 11, Folder 7), Georgetown University, Washington, DC.

Collins, Patrick D. 1976. Letter to the National Commission, November 22. The National Commission for the Protection of Human Subjects of Biomedical and Behavioral Research, 1974–1978, Bioethics Research Library (Box 11, Folder 7), Georgetown University, Washington, DC.

Collins, William B. 1968. "Lab Testing Program Tied to Prison Sex Corruption." *Philadelphia Inquirer,* September 28. *Philadelphia Inquirer* collections, Special Collections Research Center (K 6-479, 6A 44), Temple University, Philadelphia, PA.

Comfort, Nathaniel. 2009. "The Prisoner as Model Organism: Malaria Research at Stateville Penitentiary." *Studies in History and Philosophy of Science*, 40(3), 190–203.

Comics Grinder. 2014. "Review: Ghost Stalkers: Holmesburg Prison, Tonight, November 6, Thursdays at 10/9c on Destination America." [Web blog comments]. https://comicsgrinder.com/

Connor, Steven. 2004. *The Book of Skin*. London: Reaktion Books.

Coole, Diana, and Samantha Frost. 2010. *New Materialisms: Ontology, Agency, and Politics*. Durham, NC: Duke University Press.

Cooper, Melinda, and Catherine Waldby. 2014. *Clinical Labor: Tissue Donors and Research Subjects in the Global Bioeconomy*. Durham, NC: Duke University Press.

Copeland, Huey, and Krista Thompson. 2011. "Perpetual Returns: New World Slavery and the Matter of the Visual." *Representations*, 113(1), 1–15.

Crary, Jonathan. 1988. "Modernizing Vision." In Hal Foster, ed., *Vision and Visuality* (pp. 29–50). Seattle, WA: Bay Press.

Dale, Maryclaire. 2000. "Phila. Inmates Sue over Testing." *ABC News*, October 19. http://abcnews.go.com/

Daston, Lorraine, and Peter Gallison. 1992. "The Images of Objectivity." *Representations*, 0(40), 81–128.

Davis, Alan J. 1968. "Sexual Assaults in the Philadelphia Prison System and Sheriff's Vans." *Trans-action*, 6(2), 8–17.

Davis, Angela. 1998. "Racialized Punishment and Prison Abolition." In Joy James, ed., *The Angela Y. Davis Reader* (pp. 96–107). Oxford: Blackwell.

———. 2003. *Are Prisons Obsolete?* New York: Seven Stories.

Dayan, Colin. 2001. "Legal Slaves and Civil Bodies." *Nepantla: Views from South*, 2(1), 3–39.

Derrida, Jacques. 1994. "Spectres of Marx." *New Left Review*, 1(205), 31–58.

———. 2016. *Of Grammatology*. Gayatri Spivak, trans. Baltimore, MD: Johns Hopkins University Press. (Original work published 1974)

Dunn, William. 1975. "'Guinea Pigs' Taking Part in Tests at Jackson Prison Oppose Banning Program." *Detroit News*, November 17, 3A, 6A. The National Commission for the Protection of Human Subjects of Biomedical and Behavioral Research, 1974–1978, Bioethics Research Library (Box 4, Folder 5), Georgetown University, Washington, DC.

Duster, Troy. 2003. *Backdoor to Eugenics*. New York: Routledge.

Dyer, Richard. 1997. *White: Essays on Race and Culture*. New York: Routledge.

Easton, Ronald. 1975. Letter to the National Commission, November 10. The National Commission for the Protection of Human Subjects of Biomedical and Behavioral Research, 1974–1978, Bioethics Research Library (Box 4, Folder 4), Georgetown University, Washington, DC.

Edelman, Lee. 2004. *No Future: Queer Theory and the Death Drive*. Durham, NC: Duke University Press.

Eisenstein, Elizabeth. 1983. *The Printing Revolution in Early, Modern Europe*. Cambridge, UK: Cambridge University Press.

Epstein, Aaron. 1973. "Panel Rules Holmesburg 'Cruel,' Unfit." *Philadelphia Inquirer,* September 5. *Philadelphia Inquirer* collections, Special Collections Research Center (Holmesburg Prison 1968–1974), Temple University, Philadelphia, PA.

———. 1979. "At Holmesburg Prison, 320 Human Guinea Pigs." *Philadelphia Inquirer,* November 25. *Philadelphia Inquirer* collections, Special Collections Research Center (K 6–479, 6A 44), Temple University, Philadelphia, PA.

Epstein, Edgar. 2012. "Patricia Gómez + María Jesús González." *Art Papers,* 52, March/April.

Epstein, Steven. 2007. *Inclusion: The Politics of Difference in Medical Research.* Chicago: University of Chicago Press.

Fanon, Frantz. 2004. *The Wretched of the Earth.* New York: Grove.

———. 2008. *Black Skin, White Masks.* Richard Philcox, trans. New York: Grove. (Original work published 1952)

Fleetwood, Nicole. 2020. *Marking Time: Art in the Age of Mass Incarceration.* Cambridge, MA: Harvard University Press.

Foucault, Michel. 1980. *Power/Knowledge: Selected Interviews and Other Writings, 1972–1977.* Colin Gordon, ed. New York: Pantheon.

———. 1990. *The History of Sexuality: An Introduction.* New York: Vintage. (Original work published 1976)

———. 1994. *The Birth of the Clinic: An Archaeology of Medical Perception.* New York: Vintage. (Original work published 1973)

———. 1995. *Discipline and Punish: The Birth of the Prison.* New York: Vintage. (Original work published 1977)

———. 1991. "Governmentality." In Graham Burchell, Colin Gordon, and Peter Miller, eds., *The Foucault Effect: Studies in Governmentality* (pp. 87–104). Chicago: University of Chicago Press.

Foucault, Michel, and John. K Simon. 1991. "Michel Foucault on Attica: An Interview." *Social Justice,* 18(3), 26–34.

Frosch, Peter J., and Albert M. Kligman. 1976. "The Chamber-Scarification Test for Irritancy." *Contact Dermatitis,* 2, 314–324.

———. 1979. "The Duhring Chamber." *Contact Dermatitis,* 5(2), 73–81.

Fullwiley, Duana. 2007. "The Molecularization of Race: Institutionalizing Human Difference in Pharmacogenetics Practice." *Science as Culture,* 16(1), 1–30.

Gellene, Denise. 2010. "Dr. Albert M. Kligman, Dermatologist, Dies at 93." *New York Times,* February 22. https://www.nytimes.com/

Gilbert, Alexis. 2000. "Former Inmates Sue Penn over Experiments: The Plaintiffs Allege That Penn Exposed Them to Diseases." *Daily Pennsylvania,* September 20. https://www.thedp.com/

Gilmore, Ruth. 2007. *Golden Gulag: Prisons, Surplus, Crisis, and Opposition in Globalizing California.* Berkeley: University of California Press.

Ginsberg, Robert. 2004. *The Aesthetics of Ruins.* New York: Rodopi.

Gómez, Patricia, and María Jesús González. 2011a. "Doing Time: Captured Bibliography." *Philagrafika.* http://www.philagrafika.org/

———. 2011b. "Interview with Gómez + González." In Patricia Gómez and María Jesús González, eds., *Depth of Surface* (pp. 14–15). Philadelphia: Philagrafika.

Goodwin, Charles. 1994. "Professional Vision." *American Anthropologist, 96*(3), 606–633.

Gordon, Avery. 2008. *Ghostly Matters: Haunting and the Sociological Imagination.* Minneapolis: University of Minnesota Press.

Gordon, Harmony Y. 1971. "Judge to Interview Inmates Jailed in Default of Fines." *Philadelphia Inquirer,* March 12. *Philadelphia Inquirer* collections, Special Collections Research Center (scrc 171), Temple University, Philadelphia, PA.

Gordon, Lewis. 1995. *Fanon and the Crisis of the European Man: An Essay on Philosophy and the Human Sciences.* New York: Routledge.

Gorovitz, Samuel. 1976. "Research on Captive Populations." *Journal of Pharmaceutical Sciences, 65*(5), 4.

Gould, Stephen J. 1981. *The Mismeasure of Man.* New York: W. W. Norton.

Hammonds, Evelynn M. 1997. "New Technologies of Race." In Jennifer Terry and Melodie Calvert, eds., *Processed Lives: Gender and Technology in Everyday Life* (pp. 107–122), London: Routledge.

Haraway, Donna. 1988. "Situated Knowledge: The Science Question in Feminism and the Privilege of Partial Perspective." *Feminist Studies, 14*(3), 575–599.

———. 1989. *Primate Visions: Gender, Race, and Nature in the World of Modern Science.* New York: Routledge.

———. 1991. "A Cyborg Manifesto: Science, Technology, and Socialist-Feminism in the Late Twentieth Century." In Donna Haraway, ed., *Simians, Cyborgs, and Women: The Reinvention of Nature* (pp. 149–181). New York: Routledge. (Original work published 1985)

Harding, Sandra. 2006. *Science and Social Inequality: Feminist and Postcolonial Issues.* Urbana: University of Illinois Press.

———. 2011. *The Postcolonial Science and Technology Studies Reader.* Durham, NC: Duke University Press.

Hardt, Michael. 1997. "Prison Time." *Yale French Studies, 91,* 64–79.

Harkness, Jon M. 1996. "Nuremberg and the Issue of Wartime Experiments on U.S. Prisoners." *Journal of American Medical Association, 276*(20), 1672–1675.

Harris, Cheryl I. 1993. "Whiteness as Property." *Harvard Law Review, 106*(8), 1707–1791.

Hartman, Saidiya. 1997. *Scenes of Subjection: Terror, Slavery, and Self-Making in Nineteenth-Century America.* New York: Oxford University Press.

———. 2007. *Lose Your Mother: A Journey along the Atlantic Slave Route.* New York: Farrar, Straus & Giroux.

Hartouni, Val. 1997. *Cultural Conceptions: On Reproductive Technologies and the Remaking of Life.* Minneapolis: University of Minnesota Press.

Hatch, Anthony R. 2019. *Silent Cells: The Secret Drugging of Captive America.* Minneapolis: University of Minnesota Press.

Hell, Julia, and Andreas Schönle. 2009. *Ruins of Modernity.* Durham, NC: Duke University Press.

Hirsh, Jennie. 2011. "Borderlines." In Patricia Gómez and María Jesús González, eds., *Depth of Surface* (pp. 3–12). Philadelphia: Philagrafika.

Holland, Sharon P. 2000. *Raising the Dead: Readings of Death and (Black) Subjectivity.* Durham, NC: Duke University Press.

Hollister, Leo E. 1976. Letter to Philip Handler, April 9. The National Commission for the Protection of Human Subjects of Biomedical and Behavioral Research, 1974–1978, Bioethics Research Library (Box 8, Folder 13), Georgetown University, Washington, DC.

Holloway, Karla. 2011. *Private Bodies, Public Texts.* Durham, NC: Duke University Press.

Hornblum, Allen M. 1998. *Acres of Skin: Human Experiments at Holmesburg prison.* New York: Routledge.

Huyssen, Andreas. 2009. "Authentic Ruins: Products of Modernity." In Julia Hell and Andreas Schönle, eds., *Ruins of Modernity* (pp. 17–28). Durham, NC: Duke University Press.

Ihde, Don. 1983. *Existential Technics.* Albany: State University of New York Press.

Institute of Society, Ethics, and the Life Sciences. 1977. *Hastings Center Report* (February), vol. 7, no. 1. The National Commission for the Protection of Human Subjects of Biomedical and Behavioral Research, 1974–1978, Bioethics Research Library (Meeting 29, Box 15), Georgetown University, Washington, DC.

Isbister, James D. 1976. Letter to Kenneth J. Ryan, April 2. The National Commission for the Protection of Human Subjects of Biomedical and Behavioral Research, 1974–1978, Bioethics Research Library (Box 7, Folder 21), Georgetown University, Washington, DC.

Jablonski, Nina. 2012. *Living Color: The Biological and Social Meaning of Skin Color.* Berkeley, CA: University of California Press.

Jackson, Cassandra. 2011. "Visualizing Slavery: Photography and the Disabled Subject in the Art of Carrie Mae Weems." In Christopher Bell, ed., *Blackness and Disability: Critical Examinations and Cultural Interventions* (pp. 31–46). East Lansing: Michigan State University Press.

Jackson, George. 1994. *Soledad Brother: The Prison Letters of George Jackson.* Chicago: Lawrence Hill. (Original work published 1970)

Jackson, Zakiyyah I. 2020. *Becoming Human: Matter and Meaning in an Antiblack World.* New York: New York University Press.

James, Joy. 1996. *Resisting State Violence: Radicalism, Gender, and Race in U.S. Culture.* Minneapolis: University of Minnesota Press.

———, ed. 2007. *Warfare in the American Homeland: Policing and Prisons in the Penal Democracy.* Durham, NC: Duke University.

JanMohamed, Abdul R. 2005. *The Death-Bound-Subject: Richard Wright's Archaeology of Death.* Durham, NC: Duke University Press.

Johnson, Jenell. 2014. *American Lobotomy: A Rhetorical History.* Ann Arbor: University of Michigan Press.

Johnson, Tyree. 1975. "Convicts Here Used in Tests." *Philadelphia Inquirer,* July 31. *Philadelphia Inquirer* collections, Special Collections Research Center (scrc 171), Temple University, Philadelphia, PA.

Johnston, Norman B. 2000. *Forms of Constraint: A History of Prison Architecture.* Urbana: University of Illinois Press.

Johnston, Norman B., Kenneth Finkel, and Jeffrey A. Cohen. 1994. *Eastern State Penitentiary: A Crucible of Good Intentions.* Philadelphia: Philadelphia Museum of Art.

Jonsen, Albert R. 1998. *The Birth of Bioethics.* New York: Oxford University Press.

———. 2000. *A Short History of Medical Ethics.* New York: Oxford University Press.

Katz, Adolph. 1966. "Prisoners Volunteer to Save Lives: Holmesburg Inmates Test Medicines, Serve as Aides in Laboratory Work." *Philadelphia Bulletin,* February 27, 1, 6. *Philadelphia Bulletin* collections, Special Collections Research Center (579-S), Temple University, Philadelphia, PA.

Keeling, Kara. 2003. "'In the Interval': Frantz Fanon and the 'Problems' of Visual Representation." *Qui Parle, 13*(2), 91–117.

———. 2007. *The Witch's Flight: The Cinematic, the Black Femme, and the Image of Common Sense.* Durham, NC: Duke University Press.

Kelley, Robin D. G. 2002. *Freedom Dreams: The Black Radical Imagination.* Boston: Beacon Press.

Kligman, Albert M. 1966a. "Identification of Contact Allergens by Human Assay. II. Factors Influencing the Induction and Measurement of Allergic Contact Dermatitis." *Journal of Investigative Dermatology, 47*(5), 375–392.

———. 1966b. "Identification of Contact Allergens by Human Assay. III. The Maximization Test: A Procedure for Screening and Rating Contact Sensitizers." *Journal of Investigative Dermatology, 47*(5), 393–409.

———. 1974. "Sustained Protection against Superficial Bacterial and Fungal Infection by Topical Treatment: Final Comprehensive Report." Fort Detrick, MD: US Army Medical Research and Development Command.

———. 1991. "The Invisible Dermatoses." *Archives of Dermatology, 127,* 1375–1382.

Kligman, Albert M., and Enno Christophers. 1963. "Preparation of Isolated Sheets of Human Stratum Corneum." *Archives of Dermatology, 88*(6), 702–705.

Kligman, Albert M., and William Epstein. 1975. "Updating the Maximization Test for Identifying Contact Allergens." *Contact Dermatitis, 1,* 231–239.

Kligman, Albert M., and Walter B. Shelley. 1958. "An Investigation of the Biology of the Human Sebaceous Gland." *Journal of Investigative Dermatology, 30*(3), 99–125.

Kligman, Albert M., and Isaac Willis. 1975. "A New Formula for Depigmenting Human Skin." *Archives of Dermatology, 111*(1), 40–48.

Kligman, Albert M., and W. M. Wooding. 1967. "A Method for the Measurement and Evaluation of Irritants on Human skin." *Journal of Investigative Dermatology, 49*(1), 78–94.

Knorr-Cetina, Karin. 1981. *The Manufacture of Knowledge: An Essay on the Constructivist and Contextual Nature of Science.* Elmsford, NY: Pergamon Press.

Kourelis, Kostis. 2011. "Splitting Architectural Time: Gómez + González Holmesburg Prison Project." *Philagrafika*. http://www.philagrafika.org/

Kropotkin, Peter. 1998. *Fields, Factories, and Workshops*. London: Freedom Press. (Original work published 1974)

Kuhn, Thomas S. 1970. *The Structure of Scientific Revolutions*. Chicago: University of Chicago Press. (Original work published 1962)

Lacan, Jacques. 1998. *The Four Fundamental Concepts of Psychoanalysis: The Seminar of Jacques Lacan, Book XI*. Alan Sheridan, trans. New York: W. W. Norton. (Original work published 1973)

———. 2006. *Écrits*. Bruce Fink, trans. New York: W. W. Norton. (Original work published 1966)

Lafrance, Marc. 2018. "Skin Studies: Past, Present and Future." *Body & Society*, 24(1–2), 3–32.

Lala, Elisa. 2016. "City Retracts Plan to Use Shuttered Holmesburg Prison to Hold Arrested Protesters during DNC." *Philly Voice,* June 23. http://www.phillyvoice.com/

Lathers, Forest G. 1975. Letter to the National Commission, November 10. The National Commission for the Protection of Human Subjects of Biomedical and Behavioral Research, 1974–1978, Bioethics Research Library (Box 4, Folder 4), Georgetown University, Washington, DC.

Latour, Bruno. 1987. *Science in Action: How to Follow Scientists and Engineers through Society*. Cambridge, MA: Harvard University Press.

———. 1988. *The Pasteurization of France*. Cambridge, MA: Harvard University Press.

———. 1992. "Where Are the Missing Masses? The Sociology of a Few Mundane Artifacts." In Wiebe E. Bijker and John Law, eds., *Shaping Technology/Building Society* (pp. 225–258). Cambridge, MA: MIT Press.

Latour, Bruno, and Steve Woolgar. 1979. *Laboratory Life: The Construction of Scientific Facts*. Princeton, NJ: Princeton University Press.

Lawson, Allen. 1976. Testimony of the Prisoners' Rights Council before the National Commission, January 9. The National Commission for the Protection of Human Subjects of Biomedical and Behavioral Research, 1974–1978, Bioethics Research Library (Box 4, Meeting 14), Georgetown University, Washington, DC.

Lester, Jenna C., and Susan C. Taylor. 2021. "Resisting Racism in Dermatology: A Call to Action." *JAMA Dermatology*, 157(3), 267–268.

Leyden, James J. 1991. "Albert Kligman—Master of Dermatology." *Archives of Dermatology*, 127, 1392.

Leyden, James J., and Albert M. Kligman. 1978. "Interdigital Athlete's Foot: The Interaction of Dermatophytes and Resident Bacteria." *Archives of Dermatology*, 114, 1466–1472.

Lynch, Michael. 1985. *Art and Artifact in Laboratory Science: A Study of Shop Work and Shop Talk in a Research Laboratory*. London: Routledge.

———. 1988. "The Externalized Retina: Selection and Mathematization in the Visual Documentation of Objects in the Life Sciences." *Human Studies*, 11(2–3), 201–234.

Machiavelli, Niccolò. 1976. *The Prince*. James B. Atkinson, trans. Indianapolis, IN: Hacket.

Maibach, Howard I., and Albert M. Kligman. 1962. "The Biology of Experimental Human Cutaneous Moniliasis (Candida Albicans)." *Archives of Dermatology, 85*, 113–137.

Malone, Noreen. 2011. "The Case against Economic Disaster Porn." *The New Republic*, January 21, 2011. https://newrepublic.com/

Marks, Laura U. 2000. *The Skin of the Film: Intercultural Cinema, Embodiment, and the Senses*. Durham, NC: Duke University Press.

Marriott, David. 2007. *Haunted Life: Visual Culture and Black Modernity*. New Brunswick, NJ: Rutgers University Press.

Martinot, Steve, and Jared Sexton. 2003. "The Avant-Garde of White Supremacy." *Social Identities, 9*(2), 169–181.

Maugh, Thomas H., II. 2010. "Albert M. Kligman Dies at 93; Dermatologist Developed Acne, Wrinkle Treatments and Experimented on Prisoners." *Los Angeles Times*, February 24. http://articles.latimes.com/

McGuire, Danielle L. 2010. *At the Dark End of the Street, Black Women, Rape, and Resistance: A New History of the Civil Rights Movement from Rosa Parks to the Rise of Black Power*. New York: Vintage Press.

McKittrick, K. 2006. *Demonic Grounds: Black Women and the Cartographies of Struggle*. Minneapolis: University of Minnesota Press.

———. 2011. "On Plantations, Prisons, and a Black Sense of Place." *Social & Cultural Geography, 12*(8), 947–963.

McMahon, F. G. 1976. Letter to Kenneth J. Ryan, April 22. The National Commission for the Protection of Human Subjects of Biomedical and Behavioral Research, 1974–1978, Bioethics Research Library (Box 8, Folder 14), Georgetown University, Washington, DC.

Medlicott, D. 1999. "Surviving in the Time Machine: Suicidal Prisoners and the Pains of Prison Time." *Time and Society, 8*(2), 211–230.

Merton, R. K. 1938. "Motive Forces of the New Science." In R. K. Merton, ed., *Science, Technology and Society in Seventeenth Century England* (pp. 112–131), New York: Howard Fertig.

Metzl, J. 2009. *The Protest Psychosis: How Schizophrenia became a Black Disease*. Boston: Beacon Press.

Miles, T. 2015. *Tales from the Haunted South: Dark Tourism and Memories of Slavery from the Civil War Era*. Chapel Hill: University of North Carolina Press.

Mintz, M. 1973. "FDA Blacklists Penn Professor over Test Data." *Philadelphia Inquirer*, July 23. *Philadelphia Inquirer* collections, Special Collections Research Center (Phil. Inq.—Prison—1952–1965), Temple University, Philadelphia, PA.

Mirzoeff, Nick. 2011. "The Right to Look." *Critical Inquiry, 37*(3), 473–496.

Mitchell, W. J. T. 1994. *Picture Theory*. Chicago: University of Chicago Press.

———. 2002. "Showing Seeing: A Critique of Visual Culture." *Journal of Visual Culture, 1*(2), 165–181.

———. 2005. *What Do Pictures Want?: The Lives and Loves of Images*. Chicago: University of Chicago Press.

Mitford, Jessica. 1973. *Kind and Usual Punishment: The Prison Business*. New York: Knopf.

Mol, Annemarie. 2002. *The Body Multiple: Ontology in Medical Practice*. Durham, NC: Duke University Press.

Montgomery, Charles. 1971. "Cops Beef Forces at 3 City Prisons." *Philadelphia Inquirer,* September 7. *Philadelphia Inquirer* collections, Special Collections Research Center (Phil.InquirerClippingsprisonersk7-382 7-383), Temple University, Philadelphia, PA.

Mumford, Lewis. 1964. "Authoritarian and Democratic Technics." *Technology and Culture,* 5(1), 1–8.

———. 1970. *The Myth of the Machine: The Pentagon of Power*. New York: Harcourt Brace Jovanovich.

Muñoz, José E. 2009. *Cruising Utopia: The Then and There of Queer Futurity*. New York: New York University Press.

Naedele, Walter F. 2010. "Albert M. Kligman, 93, Dermatology Researcher." *Philadelphia Inquirer,* February 21, 2010. https://www.inquirer.com/

Nast, Heidi, and Steve Pile, eds. 1998. *Places through the Body*. London: Routledge.

National Commission for the Protection of Human Subjects of Biomedical and Behavior Research. 1975, July 26. "Recapitulation of Impressions Following Site Visit to the State Prison of Southern Michigan at Jackson." The National Commission for the Protection of Human Subjects of Biomedical and Behavioral Research Collections, 1974–1978, Bioethics Research Library (Box 25, Report no. NCPHS/M-76/01), Georgetown University, Washington, DC.

———. 1975, November 11. Meeting minutes. The National Commission for the Protection of Human Subjects of Biomedical and Behavioral Research Collections, 1974–1978, Bioethics Research Library (Box 4), Georgetown University, Washington, DC.

———. 1976. "Report and Recommendations: Research Involving Prisoners." *Department of Health, Education, and Welfare*. Publication No. (OS) 76–131.

———. 1979. "The Belmont Report: Ethical Principles and Guidelines for the Protection of Human Subjects of Research." Department of Health, Education, and Welfare. https://www.hhs.gov/

National Prison Project. 1976. Testimony of NPP of the American Civil Liberties Union for the National Commission. The National Commission for the Protection of Human Subjects of Biomedical and Behavioral Research, 1974–1978, Bioethics Research Library (Box 4, Meeting 14), Georgetown University, Washington, DC.

Nelson, Alondra. 2011. *Body and Soul: The Black Panther Party and the Fight against Medical Discrimination*. Minneapolis: University of Minnesota Press.

Nelson, Peggy. 2010. "Ultimate Ruin Porn." *HiLowBrow,* December 17. http://hilobrow.com/

New York Times. 1975, October 2. "H.E.W. Official Opposes Ban on Research with Prisoners." p. C-20. The National Commission for the Protection of Human Subjects of

Biomedical and Behavioral Research, 1974–1978, Bioethics Research Library (Meeting 11, Box 3), Georgetown University, Washington, DC.

Norris, David A. 1987. "Book Review. The Tradition of Excellence: Dermatology at the University of Pennsylvania, 1870–1985." *Journal of Investigative Dermatology, 88,* 625.

Pallasmaa, Juhani. 2005. *The Eyes of the Skin: Architecture and the Senses.* Hoboken, NJ: John Wiley & Sons.

Patterson, Orlando. 1982. *Slavery and Social Death: A Comparative Study.* Cambridge, MA: Harvard University Press.

Pennsylvania Prison Society. 1912. "The Philadelphia County Prison at Holmesburg." *Journal of Prison Discipline and Philanthropy, 51,* 19–21.

Perlman, Katherine L., Elizabeth L. Klein, Joyce H. Park. 2020. "Racial Disparities in Dermatology Training: The Impact on Black Patients." *Cutis, 106*(6), 300–301.

Petechsky, Rosalind P. 1987. "The Power of Visual Culture in the Politics of Reproduction." *Feminist Studies, 13*(2), 263–292.

Petryna, Adriana. 2007. "Clinical Trials Offshored: On Private Sector Science and Public Health." *BioSocieties, 2*(1), 21–40.

———. 2009. *When Experiments Travel: Clinical Trials and the Global Search for Human Subjects.* Princeton, NJ: Princeton University Press.

Pharmaceutical Manufacturer's Association. 1975, January 9. Proceedings for the National Commission. The National Commission for the Protection of Human Subjects of Biomedical and Behavioral Research, 1974–1978, Bioethics Research Library, Georgetown University, Washington, DC.

———. 1975, February 11. "Statement of Principles on the Conduct of Pharmaceutical Research in the Prison Environment." The National Commission for the Protection of Human Subjects of Biomedical and Behavioral Research, 1974–1978 (Box 1, Folder 8), Bioethics Research Library, Georgetown University, Washington, DC.

Philadelphia Inquirer. 1938, August 20. "600 Prisoners on Food Strike at Holmesburg." *Philadelphia Inquirer* collections, Special Collections Research Center (K 7a-33 7m-29, Homesburg County Prison—convicts—1938–1969), Temple University, Philadelphia, PA.

———. 1954, March. "Prisoners Stage Hunger Strike." *Philadelphia Inquirer* collections, Special Collections Research Center (K 7a-33 7m-29, Homesburg County Prison—convicts—1938–1969), Temple University, Philadelphia, PA.

———. 1969a, June 3. "Prison Fast Called 'Legitimate' by Official." *Philadelphia Inquirer* collections, Special Collections Research Center (K 7a-33 7m-29, Homesburg County Prison—convicts—1938–1969), Temple University, Philadelphia, PA.

———. 1969b, June 3. "Convicts End Hunger Strike." *Philadelphia Inquirer* collections, Special Collections Research Center (K 7a-33 7m-29, Homesburg County Prison—convicts—1938–1969), Temple University, Philadelphia, PA.

———. 1969c, October 21. "Chaplain Calls Phila. Prisons 'Classic Example of White Racism.'" *Philadelphia Inquirer* collections, Special Collections Research Center

(K 7a-33 7m-29, Homesburg County Prison—convicts—1938–1969), Temple University, Philadelphia, PA.

———. 1971a, January. "Speedy Justice for the Accused." *Philadelphia Inquirer* collections, Special Collections Research Center (scrc 171), Temple University, Philadelphia, PA.

———. 1971b, September 7. "Holmesburg: Often 3 in a Cell." *Philadelphia Inquirer* collections, Special Collections Research Center (K 7a-33 7m-29, Holmesburg County Prison—convicts—gen.—1971), Temple University, Philadelphia, PA.

Pile, Steve. 1996. *The Body and the City: Psychoanalysis, Space and Subjectivity*. London: Routledge.

Plewig, Gerd. 1991. "A Giant in the Field: Albert Montgomery Kligman." *Archives of Dermatology*, 127(9), 1415–1416.

Pollock, Anne. 2012. *Medicating Race: Heart Disease and Durable Preoccupations with Difference*. Durham, NC: Duke University Press.

Prosser, Jay. 2001. "Skin Memories." In Sara Ahmed and Jackie Stacey, eds., *Thinking through the Skin* (pp. 52–68). New York: Routledge.

Raengo, Alessandra. 2013. *On the Sleeve of the Visual: Race as Face Value*. New Lebanon, NH: Dartmouth College Press.

Raiford, Leigh. 2011. *Imprisoned in a Luminous Glare: Photography and the African American Freedom Struggle*. Chapel Hill: University of North Carolina Press.

Rebora, Alfredo, Richard R. Marples, and Albert M. Kligman. 1963a. "Erosio Interdigitalis Blastomycetica." *Archives of Dermatology*, 108, 66–68.

———. 1963b. "Experimental Infection with Candida Albicans." *Archives of Dermatology*, 108, 69–73.

Reiter, Keramet. 2009. "Experimentation on Prisoners: Persistent Dilemmas in Rights and Regulations." *California Law Review*, 97(2), 501–566.

Richards, Bill. 1975. "Army May Drop Testing Funds." *Washington Post*, September 25. The National Commission for the Protection of Human Subjects of Biomedical and Behavioral Research, 1974–1978, Bioethics Research Library (Box 3, Meeting 11), Georgetown University, Washington, DC.

Robbins, William. 1983. "Dioxin Tests Conducted in the '60s on 70 Philadelphia Inmates, Now Unknown." *New York Times*, July 17. https://www.nytimes.com/

Roberts, Dorothy. 2011. *Fatal Invention: How Science, Politics, and Big Business Recreate Race in the Twenty-First Century*. New York: New Press.

Robertson, Patricia S. 2011. "Printing the Past: Gómez + González' Monoprints." In Patricia Gómez and María Jesús González, eds., *Depth of Surface* (pp. 18, 25). Philadelphia: Philagrafika.

Roca, José. 2011. "Doing Time/Depth of Surface." *Philagrafika*. http://www.philagrafika.org/

Rodríguez, Dylan. 2006. *Forced Passages: Imprisoned Radical Intellectuals and the U.S. Prison Regime*. Minneapolis: University of Minnesota Press.

———. 2007. "Forced Passages." In Joy James, ed., *Warfare in the American Homeland: Policing and Prison in the Penal Democracy* (pp. 36–57). Durham, NC: Duke University.

Rogers, Molly, and David W. 2010. *Delia's Tears: Race, Science, and Photography in Nineteenth-Century America.* New Haven, CT: Yale University Press.

Roma, Thomas. 2005. *In Prison Air: The Cells of Holmesburg Prison.* Brooklyn, NY: Powerhouse Books.

Rose, Gillian. 1993. *Feminism and Geography: The Limits of Geographical Knowledge.* Minneapolis: University of Minnesota Press.

Rose, Nikolas. 2007. *The Politics of Life Itself: Biomedicine, Power, and Subjectivity in the Twentieth Century.* Princeton, NJ: Princeton University Press.

Rosenberg, Amy S. 2012. "Bits of Holmesburg Prison's History Preserved for Moore College Exhibit." *Philly,* February 2, 2012. http://articles.philly.com/

Rosenfeld, Stephan. 1977. "City Fined in Snub of Jail Edict." *Philadelphia Inquirer,* December 1. *Philadelphia Inquirer* collections, Special Collections Research Center (Holmesburg Prison Conditions 1975-), Temple University, Philadelphia, PA.

Rossi, Beth, and Elizabeth Rossi. 2003. "Research Subjects Protest Prof: Dermatologist Honored for Achievement, Accused of Unethical Experimentation." *Daily Pennsylvanian,* October 30. http://www.thedp.com/

Rubin, Mike. 2011. "Capturing the Idling of the Motor City." *New York Times,* August 18. http://www.nytimes.com/

Ryan, Kenneth J. 1975. Letter to William G. Milliken, December 4. The National Commission for the Protection of Human Subjects of Biomedical and Behavioral Research, 1974–1978, Bioethics Research Library (Box 4, Folder 3), Georgetown University, Washington, DC.

Rykwert, Joseph. 1996. *The Dancing Column: On Order in Architecture.* Cambridge, MA: MIT Press.

Sabo, Donald F., Terry Allen Kupers, and Willie James London. 2001. Prison Masculinities. Philadelphia, PA: Temple University Press.

Scandura, Jani. 2007. *Down in the Dumps: Place, Modernity, American Depression.* Durham, NC: Duke University Press.

Scarry, Elaine. 1985. *The Body in Pain: The Making and Unmaking of the World.* New York: Oxford University Press.

Schmeck, Harold M. 1976a. "U.S. Panel Weighs Tests on Inmates." *New York Times,* March 14. The National Commission for the Protection of Human Subjects of Biomedical and Behavioral Research, 1974–1978, Bioethics Research Library (Box 17, Folder 22), Georgetown University, Washington, DC.

———. 1976b. "Report Urges Ban on Prison Tests." *New York Times,* June 11. The National Commission for the Protection of Human Subjects of Biomedical and Behavioral Research, 1974–1978, Bioethics Research Library (Box 9, Folder 6), Georgetown University, Washington, DC.

Sekula, Allan. 1986. "The Body and the Archive." *October, 39,* 3–64.

Seltzer, Mark. 1997. "Wound Culture: Trauma and the Pathological Public Sphere." *October, 80,* 3–26.

Serres, Michael. 2013. "Betrayal: The Thanatocracy." *Public,* 24(48), 19–40. (Original work published 1974)

Shakur, Assata. 1988. *Assata: An Autobiography*. London: Zed Books.

Shanks, Michael, David Platt, and William Rathje. (2004). "The Perfume of Garbage: Modernity and the Archaeological." *Modernism|modernity*, 11(1), 61–83.

Sharpe, Jenny. 2003. *Ghosts of Slavery: A Literary Archaeology of Black Women's Lives*. Minneapolis: University of Minnesota Press.

Shuttleworth, Ken. 1971. "DA Seeks Court Order to Force Prison Reforms." *Philadelphia Inquirer*, February. *Philadelphia Inquirer* collections, Special Collections Research Center (scrc 171), Temple University, Philadelphia, PA.

Siebers, Tobin. 2010. *Disability Aesthetics*. Ann Arbor: University of Michigan Press.

Smith, Cedric M. 1976. Letter to Robert W. Kastenmeier, March 4. The National Commission for the Protection of Human Subjects of Biomedical and Behavioral Research, 1974–1978, Bioethics Research Library (Box 8, Folder 10), Georgetown University, Washington, DC.

Smith, Robert J., and Brittany U. Oliver. 2021. "Advocating for Black Lives—A Call to Dermatologists to Dismantle Institutionalized Racism and Address Racial Health Inequities." *JAMA Dermatology*, 157(2), 155–156.

Smith, Shawn M. 2004. *Photography on the Color Line: W. E .B. Du Bois, Race, and Visual Culture*. Durham, NC: Duke University Press.

———. 2013. *At the Edge of Sight: Photography and the Unseen*. Durham, NC: Duke University Press.

Sobchack, Vivian. 1992. *The Address of the Eye: A Phenomenology of Film Experience*. Princeton, NJ: Princeton University Press.

Sontag, Susan. 2003. *Regarding the Pain of Others*. New York: Picador.

Spillers, Hortense. 2003. "Mama's Baby, Papa's Maybe: An American Grammar Book." In Hortense Spillers, ed., *Black, White and in Color: Essays on American Literature and Culture* (pp. 203–229). Chicago: University of Chicago Press. (Original work published 1987)

Spivak, Gayatri. C. 1995. "Ghostwriting." *Diacritics*, 25(2), 64–84.

Stanley, Eric A., and Nat Smith. 2015. *Captive Genders: Trans Embodiment and the Prison Industrial Complex*. Edinburgh: AK Press.

Steele, Jonathan. 1977. "U.S. to Keep Testing Drugs on Prisoners." *Washington Post*, August 27. The National Commission for the Protection of Human Subjects of Biomedical and Behavioral Research, 1974–1978, Bioethics Research Library (Box 17, Folder 8), Georgetown University, Washington, DC.

Stepan, Nancy L. 2001. *Picturing Tropical Nature*. Ithaca, NY: Cornell University Press.

Stephens, Michelle A. 2014. *Skin Acts: Race, Psychoanalysis, and the Black Male Performer*. Durham, NC: Duke University Press.

Stobbe, Michael. 2011. "'First Do No Harm': Past Medical Testing on Americans Revealed." *Times and Democrat*, March 11. https://thetandd.com/

Strauss, John S., William L. Epstein, Robert M. Lavker, and James J. Leyden. 1987. "Albert Montgomery Kligman." *Journal of Investigative Dermatology*, 88(3), 1s.

Tagg, John. 1993. *The Burden of Representation: Essays on Photographies and Histories*. Minneapolis: University of Minnesota Press.

TallBear, Kim. 2013. *Native American DNA: Tribal Belonging and the False Promise of Genetic Science*. Minneapolis: University of Minnesota Press.

Tate, Shirley. 2001. "'That Is My Star of David': Skin, Abjection and Hybridity." In Sara Ahmed and Jacki Stacey, eds., *Thinking through the Skin* (pp. 209–222). New York: Routledge.

———. 2015. *Skin Bleaching in Black Atlantic Zones: Shade Shifters*. Basingstoke: Palgrave Macmillan.

Taussig, Michael. 1993. *Mimesis and Alterity: A Particular History of the Senses*. New York: Routledge.

Taylor, Diana. 2014. "Trauma in the Archive." In Elspeth Brown and Thy Phu, eds., Feeling Photography (pp. 181–203). Durham, NC: Duke University Press.

Tenney, John E.L. 2014. "'Ghost Stalkers': Episode 4—Holmesburg Prison: Notes and Thoughts." *Weird Lectures,* November 7. http://weirdlectures.com/

Teruso, Julia. 2016. "With Convention Protests Looming, City Is Readying Holmesburg Prison." *Philly,* June 22. http://www.philly.com/

Thompson, Krista. 2011. "The Evidence of Things Not Photographed: Slavery and Historical Memory in the British West Indies." *Representations, 113*(1), 39–71.

Tilley, Christopher. 1994a. *A Phenomenology of Landscape: Places, Paths, and Monuments*. Oxford: BERG.

———. 1994b. "Interpreting Material Culture." In Susan M. Pearce, ed., *Interpreting Objects and Collections* (pp. 67–75). London, UK: Routledge.

Tilley, Helen. 2011. *Africa as Living Laboratory: Empire, Development, and the Problem of Scientific Knowledge, 1870–1950*. Chicago: University of Chicago Press.

Trouillot, Michel-Rolph. 1995. *Silencing the Past: Power and the Production of History*. Boston: Beacon Press.

Tu, Thuy Linh Nguyen. 2021. *Experiments in Skin: Race and Beauty in the Shadow of Vietnam*. Durham, NC: Duke University Press.

Tuan, Yi Fu. 2001. *Space and Place: The Perspective of Experience*. Minneapolis: University of Minnesota Press.

Turner, Terence S. 2012. "The Social Skin." *Journal of Ethnographic Theory, 2*(2), 486–504.

United States Congress. (1973). "Quality of Health Care—Human Experimentation, 1973." Hearings before the Subcommittee on Health for the Committee on Labor and Public Welfare, Ninety-Third Congress.

United States National Library of Medicine. 2016. "Trends, Charts, and Maps." *Clinicaltrials.* https://clinicaltrials.gov/

Urbina, Ian. 2006. "Panel Suggests Using Inmates in Drug Trials." *New York Times,* August 13. http://www.nytimes.com/

Vadala, Nick. 2017. "'Against the Night': A Philly Ghost Story Filmed at Holmesburg Prison— Without the Ghosts." *Philadelphia Inquirer*, September 14. https://www.inquirer.com/

Visperas, Cristina M. 2019. "The Able-Bodied Slave." *Journal of Literary and Cultural Disability Studies, 13*(1), 93–110.

Wacquant, Loïc. 2002. "From Slavery to Mass Incarceration: Re-thinking the 'Race Question' in the U.S." *New Left Review*, 13, 41–60.

Wailoo, Keith. 2014. *Pain: A Political History*. Baltimore, MD: Johns Hopkins University Press.

Wald, Priscilla. 2012. "Cells, Genes, and Stories: HeLa's Journey from Labs to Literature." In Keith Wailoo, Alondra Nelson, and Catherine Lee, eds., *Genetics and the Unsettled Past: The Collision of DNA, Race, and History* (pp. 247–265). New Brunswick, NJ: Rutgers University Press.

Wallace, Michelle. 2004. *Dark Designs and Visual Culture*. Durham, NC: Duke University Press. (Original work published 1990)

Wanzo, Rebecca. 2015. *Suffering Will Not Be Televised: African American Women and Sentimental Political Storytelling*. Albany: State University of New York Press.

Waring, Tom. 2015. "Prison Construction a Hot Topic at Holmesburg Civic." *Northeast Times*, June 17. http://www.bsmphilly.com/

Warner, Michael. 1990. *The Letters of the Republic: Publication and the Public Sphere in Eighteenth Century America*. Cambridge, MA: Harvard University Press.

Washington, Harriett. 2008. *Medical Apartheid: The Dark History of Medical Experimentation on Black Americans from Colonial Times to the Present*. New York: Doubleday.

Washington Post. 1978, March 12. "Medical Research in Prisons." The National Commission for the Protection of Human Subjects of Biomedical and Behavioral Research, 1974–1978, Bioethics Research Library (Box 21, Folder 10), Georgetown University, Washington, DC.

Way, E. Leong. 1976. Letter to Kenneth J. Ryan, March 4. The National Commission for the Protection of Human Subjects of Biomedical and Behavioral Research, 1974–1978, Bioethics Research Library (Box 7, Folder 21), Georgetown University, Washington, DC.

Weheliye, Alexander G. 2014. *Habeas Viscus: Racializing Assemblages, Biopolitics, and Black Feminist Theories of the Human*. Durham, NC: Duke University Press.

Weisberg, Timothy (Writer), and Nick Groff (Director). 2014. *Ghost Stalkers* [TV series]. Silver Spring, MD: Destination America.

———. 2014. "Holmesburg Prison" (Season 1, Episode 4 [TV series episode]. In Nick Groff, (Executive Producer), *Ghost Stalkers*. Silver Spring, MD: Destination America.

Weiss, Gail. 2013. *Body Images: Embodiment as Intercorporeality*. New York: Routledge.

Welsome, Eileen. 1999. *The Plutonium Files: American's Secret Experiments in the Cold War*. New York: Delta.

Wiegman, Robyn. 1995. *American Anatomies: Theorizing Race and Gender*. Durham, NC: Duke University Press.

Wilderson, Frank B. 2010. *Red, White, and Black: Cinema and the Structure of U.S. Antagonisms*. Durham, NC: Duke University Press.

Willis, Isaac, and Albert M. Kligman. 1968a. "The Mechanism of Photoallergic Contact Dermatitis." *Journal of Investigative Dermatology*, 51(5), 378–384.

———. 1968b. "The Mechanism of the Persistent Light Reactor." *Journal of Investigative Dermatology*, 51(5), 385–394.

Wilson, Mabel O. 1998. "Dancing in the Dark: The Inscription of Blackness in Le Corbusier's Radiant City." In Heidi J. Nast and Steve Pile, eds., *Places through the Body* (pp. 99–113). London: Routledge.

Winner, Langdon. 1986. *The Whale and the Reactor: A Search for Limits in an Age of High Technology*. Chicago: University of Chicago Press.

World Medical Association. 1964. *Declaration of Helsinki*. http://www.wma.net/

Wynter, Sylvia. 1987. "On Disenchanting Discourse: 'Minority' Literary Criticism and Beyond." *Cultural Critique*, 7, 207–244.

Wyrick, Bob. 1977. "Prisoner Drug Tests Cut, Different Subjects Sought." *Newsday*, August 28. The National Commission for the Protection of Human Subjects of Biomedical and Behavioral Research, 1974–1978, Bioethics Research Library (Box 17, Folder 6), Georgetown University, Washington, DC.

Yesley, Michael S. 1976. Letter to Patrick D. Collins, December 6. The National Commission for the Protection of Human Subjects of Biomedical and Behavioral Research, 1974–1978, Bioethics Research Library (Box 11, Folder 7), Georgetown University, Washington, DC.

INDEX

Abel, Elizabeth, 77

abolition, 3–4, 26, 108, 132, 168–69, 173

absence: and blackness, 23, 24, 179n7, 184n36; and photography, 37; and prison laboratories, 16; and ruinscapes, 87, 92; and skin, 164

acne, 32, 33. *See also* tretinoin experiments

Acres of Skin (Hornblum), 15, 191n15

addiction, 14, 182n23

aesthetics, carceral, 100, 108, 196n25. *See also* artwork

Agassiz, Louis, 6–7, 181nn14–15, 183n28

agency: and architecture, 85; and artwork, 107–8; and bioethics, 128, 131–32, 150–52; of captive subjects, 54–56, 73–81, 120, 141, 147–54, 155, 192n19, 205n35; and informed consent, 129, 133–36; and photography, 77, 191n15; and right to participation, 136–37; and skin, 55, 57, 60, 76, 80, 81; and technology, 190n5; theories of, 189n3

Agent Orange, 32. *See also* psychochemicals

Ahmed, Sara, 3, 199n43

allergy experiments: and agency, 55; and black skin, 51, 53–54, 62, 69, 74; images of, 65, 72–73, 191n15; methods, 56–61, 190n9; and numerical evidence, 67, 68, 69; and pain, 55–56; and visual analysis, 61, 63–64, 66, 70, 191n14; on white skin, 69–71. *See also* skin experiments

American Civil Liberties Union Foundation (ACLU), 14–15, 133–34, 149, 161

American Correctional Association, 139

Anarcha, 7. *See also* enslavement

Anderson, Patrick, 120

anesthesia, 7–8

animal experimentation, 11, 39, 188n24. *See also* bioethics

Anthony (Holmesburg Survivor), 162

Anzieu, Didier, 141–42, 143, 152, 203n25, 204n26, 206n36

apartheid, medical, 8. *See also* racism, scientific

applied research, 31, 32. *See also* clinical trials

architecture. *See* spaces and architecture

archives, 87, 91, 92, 100, 106, 108, 193n8, 194n15

Archives of Dermatology, 29

Army studies, 1, 14, 32, 161–64, 165, 171, 207nn6–7

artifacts. *See* photography, scientific; publications, scientific; ruins and ruinscapes

Art Papers, 107

artwork, 98–101, 105, 106–9, 196n24, 196n25, 197n32. *See also* photography

ascetism, 76, 185n1

Attica, New York, 121, 172, 199n49

audio recordings, 100–101, 105

authenticity, 99, 196n24, 197n32

Baker, Courtney, 148, 149, 152

Barad, Karen, 57

Barker, Jennifer, 189n29, 200n51

Barthes, Roland, 35, 95, 187n16, 195n18

basic research, 31. *See also* clinical trials

ABOUT THE AUTHOR

CRISTINA MEJIA VISPERAS is Assistant Professor of Communication at the University of Southern California. She received her PhD in Communication, Science Studies, at the University of California, San Diego.